番茄基因组设计育种

FANQIE JIYINZU SHEJI YUZHONG

［法］玛蒂尔德·科斯　赵建涛
［塞内加尔］伊西多尔·迪乌夫　王娇娇
［法］维罗尼克·勒费弗尔　［法］伯纳德·卡罗梅尔　著
［法］米歇尔·热纳尔　［法］纳迪亚·贝尔廷
梁　燕　张　飞　吴　浪　李云洲　刘　博　译

中国农业出版社
北　京

番茄基因组设计育种

内 容 简 介

番茄是世界上种植范围最广、消费量最大的蔬菜作物，其适应性强，在多种环境条件下均可种植。番茄种植主要有露地和设施两种形式，加工番茄以露地种植为主，而鲜食番茄主要在温室等设施条件下种植，无论何种种植形式，番茄在种植过程中都会受到来自环境的各种生物和非生物胁迫影响，当今基因组学等资源为加快番茄遗传改良的进程提供了条件。

本书将从以下几个方面介绍气候智能型番茄的基因组设计育种：第一，重点介绍智能型番茄育种在产量、品质和环境适应性方面面临的主要挑战以及气候变化对育种目标的影响；第二，介绍目前可用于智能型番茄育种的遗传和基因组资源，并就番茄的传统育种技术和分子育种技术进行比较分析，提出一个基于生理生态过程的模型以及能够实现新的理想型育种目标的重要策略；第三，介绍如何利用新的生物技术工具培育环境智能型番茄新品种。

前言
FOREWORD

生产中使用的番茄品种主要有有限生长型和无限生长型两种类型。加工番茄大多属于有限生长型，主要在露地种植，用于加工成各种番茄制品；鲜食番茄多为无限生长型，种植环境多种多样，露地或各种设施包括人工智能温室均可种植。

番茄（*Solanum lycopersicum* L.）是茄科家族中的重要一员。番茄为二倍体($2n=2x=24$)自花授粉作物，基因组中等大小，约为 950 Mb。番茄栽培种及其 12 个近缘野生种均起源于南美洲，相互可以杂交。世界上有几个较大的番茄种质资源基因库，保存有 70 000 多个番茄品种，除此以外，还有突变体库和分离群体库等可利用的科学研究资源。

番茄是一种长期用于遗传研究的模式植物，研究人员在番茄上发现了许多重要表型变异和突变体，很多抗病基因的图位克隆及其功能鉴定都是在番茄上完成的。番茄也是果实发育和生理研究的模式植物。由于其遗传转化比较容易，番茄成为世界上首个进行商业化生产和销售的转基因食品（Kramer et al.，1994）。

本书将从三个方面介绍智能型番茄的基因组设计育种：第一，介绍智能型番茄育种面临的主要挑战，着重介绍产量、果实品质以及环境适应性等相关的育种目标以及气候变化对这些育种目标的影响；第二，介绍智能型番茄育种可以利用的遗传和基因组资源，比较分析传统育种技术和分子育种技术，重点介绍一种生理生态模型，以及利用这种模型设计出符合育种目标的新的理想型番茄品种；第三，举例说明如何利用新的生物技术工具培育智能型番茄新品种。

由于编者水平有限，书中难免有不足之处，敬请读者批评指正。

编 者
2022 年 12 月 20 日

目 录 CONTENTS

前言

1 番茄智能型育种面临的主要挑战 ································· 1
1.1 番茄产量 ·· 1
1.2 番茄果实品质 ··· 2
1.2.1 番茄果实营养品质 ··· 2
1.2.2 番茄果实感官品质 ··· 5
1.2.3 适度胁迫对番茄果实品质的影响 ························· 7
1.3 番茄对生物胁迫的抗性及对非生物胁迫的耐受性 ······· 8
1.3.1 番茄对生物胁迫的抗性 ······································ 8
1.3.1.1 番茄病虫害的种类 ······································· 8
1.3.1.2 气候变化对番茄病虫抗性的影响 ··················· 9
1.3.1.3 番茄对新病虫害的抗性 ································ 9
1.3.2 番茄对非生物胁迫的耐受性 ······························ 10
1.3.2.1 水分胁迫 ·· 11
1.3.2.2 盐胁迫 ··· 12
1.3.2.3 温度胁迫 ·· 13
1.3.2.4 矿质营养胁迫 ·· 15
1.3.3 番茄对综合胁迫的耐受性 ·································· 17

2 番茄的遗传和基因组资源 ··· 19
2.1 番茄的遗传资源 ··· 19
2.1.1 番茄的起源及其野生种 ····································· 19
2.1.2 番茄的种质资源库 ·· 19
2.1.3 番茄的突变体库 ··· 20
2.2 番茄的分子标记、基因和 QTL ································· 22
2.2.1 番茄分子标记的发展 ··· 22
2.2.2 番茄数量性状 QTL ·· 22
2.2.3 番茄表型解析专用群体库 ·································· 24

2.2.4 番茄抗病基因与 QTL ………………………………………………… 25
      2.2.4.1 番茄抗病基因与 QTLs 资源的发掘 …………………………… 25
      2.2.4.2 番茄抗病基因与 QTLs 的染色体分布 ………………………… 30
      2.2.4.3 番茄抗病基因与 QTLs 的分子基础 …………………………… 31
  2.3 番茄的基因组资源 …………………………………………………………… 33
   2.3.1 番茄参考基因组序列 …………………………………………………… 33
   2.3.2 番茄基因组重测序 ……………………………………………………… 34
  2.4 番茄的 SNP 标记 ……………………………………………………………… 35
   2.4.1 番茄 SNP 的发现 ……………………………………………………… 35
   2.4.2 番茄 SNP 微阵列芯片 ………………………………………………… 36
   2.4.3 番茄基因型填充 ………………………………………………………… 37
  2.5 番茄的遗传多样性分析 ……………………………………………………… 38
  2.6 番茄中已克隆的基因和 QTL ………………………………………………… 39
  2.7 番茄新的基因和 QTL 鉴定 ………………………………………………… 40
  2.8 番茄全基因组关联分析 ……………………………………………………… 41
   2.8.1 番茄全基因组关联分析的条件 ………………………………………… 41
   2.8.2 番茄全基因组关联的 Meta 分析 ……………………………………… 43
  2.9 番茄非生物胁迫耐受性的遗传解析 ………………………………………… 44
   2.9.1 番茄基因型与环境互作（G×E）的遗传调控 ………………………… 44
   2.9.2 番茄嫁接对胁迫耐受性的影响 ………………………………………… 49
  2.10 番茄组学分析 ………………………………………………………………… 50
   2.10.1 番茄代谢组学分析 …………………………………………………… 50
   2.10.2 番茄转录组学分析与 eQTL 定位 …………………………………… 50
   2.10.3 番茄多组学分析 ……………………………………………………… 50
   2.10.4 番茄 MicroRNA 与表观遗传修饰 …………………………………… 51
  2.11 番茄数据库 …………………………………………………………………… 54

3 **番茄智能型基因组设计育种** ……………………………………………………… 55
  3.1 番茄传统育种 ………………………………………………………………… 55
  3.2 番茄标记辅助选择育种 ……………………………………………………… 55
   3.2.1 番茄标记辅助回交法 …………………………………………………… 56
   3.2.2 番茄 QTL 标记辅助选择 ……………………………………………… 56
   3.2.3 番茄高阶回交选择 ……………………………………………………… 57
   3.2.4 番茄聚合设计育种 ……………………………………………………… 58
   3.2.5 番茄抗病虫育种 ………………………………………………………… 58

  3.3 番茄基因组选择育种 …………………………………………………… 59

4 番茄理想型的模型设计 ………………………………………………………… 62

  4.1 番茄理想型的概念 ……………………………………………………… 62

  4.2 番茄基因型、环境、管理互作（G×E×M）的预测模型 ………… 63

  4.3 番茄理想型的模型设计 ………………………………………………… 65

  4.4 番茄模型设计育种的应用前景 ………………………………………… 67

5 番茄生物技术与基因工程 ……………………………………………………… 69

  5.1 番茄基因工程技术的发展 ……………………………………………… 69

  5.2 番茄基因工程技术 ……………………………………………………… 70

    5.2.1 番茄基因沉默与过表达技术 …………………………………… 71

    5.2.2 番茄基因组编辑技术 …………………………………………… 73

    5.2.3 番茄综合基因组工程 …………………………………………… 74

  5.3 番茄病虫抗性的基因工程 ……………………………………………… 75

  5.4 基因编辑植物的监管 …………………………………………………… 75

6 结论与展望 ………………………………………………………………………… 77

参考文献 ……………………………………………………………………………… 78

1 番茄智能型育种面临的主要挑战

确定育种目标是新品种选育的第一步。育种目标通常是由育种家根据作物生长条件的多样性、产品用途（鲜食或加工）以及销售地区的消费者喜好等确定的。番茄的主要育种目标包含以下3个方面：①产量；②品质，包括营养品质和感官品质；③抗性，包括对生物和非生物胁迫等生长环境的适应性。气候与环境变化使番茄主要育种目标的实现正面临一系列挑战。

1.1 番茄产量

世界番茄产量已经从1988年的0.64亿t增长到2017年的1.82亿t。1995年以来，中国番茄种植面积和产量迅速提高，2017年生产面积达到480万hm^2，产量涨幅达0.6亿t（图1-1），成为世界第一番茄生产大国。世界番茄产量的增加一方面是由于种植面积的增长，另一方面也归功于新品种的选育和产量的提升。

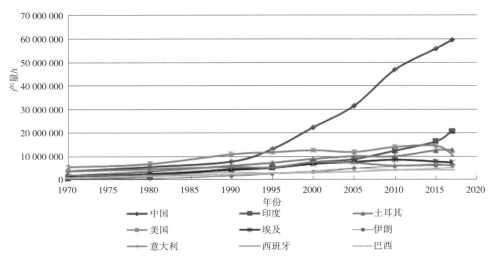

图1-1 世界上9个番茄主产国不同年份番茄产量变化

番茄的平均单产在逐年增长，由 1961 年 16t/hm² 增至目前的 37t/hm²。不同国家和地区之间的单产差异较大，南欧地区温室番茄的平均产量为 50～80t/hm²，荷兰和比利时的温室一年内可持续生产番茄 11 个月，产量超过 400t/hm²。世界番茄平均产量为 37t/hm²，荷兰达到 500t/hm²，中国为 56t/hm²。中国的番茄生产过去以露地生产为主，近年来日光温室生产得到了迅猛发展（Cao et al.，2019）。

番茄产量主要取决于品种特性和栽培条件。产量由定植株数、单株坐果数和单果重构成。从果实大小和形状来看，鲜食品种的单果重，轻者小于 20g，如樱桃番茄（*S. lycopersicum cerasiforme*），重者大于 200g，如牛心番茄。果实大小取决于开花前的细胞数量，而最终大小则主要取决于果实细胞膨大的速率和持续时间（Ho，1996）。此外，种子的数量及果实之间的竞争也会影响果实的大小（Bertin et al.，2002；Bertin et al.，2003）。果实对环境中的生物和非生物胁迫非常敏感，这些胁迫通常会导致果实脱落（Ruan et al.，2012）。果实数量受花序结构的调控，而花数的增加常会引起落花（Soyk et al.，2017a；Soyk et al.，2017b）。番茄果实的形状从扁圆形到椭圆形，有短有长，多种多样，而番茄果实的具体形状在果实的心皮发育阶段就已经确定。4 个基因位点的各类突变体就能解释大多数番茄果实的形状变异（Rodríguez et al.，2011）。

1.2 番茄果实品质

1.2.1 番茄果实营养品质

番茄具有较高的营养价值，并具有一定的保健功效，食用番茄可以降低罹患心血管疾病和一些癌症的风险（Giovannucci，1999）。番茄的营养价值与其果实中有关初生代谢和次生代谢的物质的组分有关（表 1 - 1）（Bramley，2000）。番茄中的番茄红素可以使果实呈现红色，而且也是一种膳食抗氧化剂。

表 1 - 1 番茄果实的营养组成和含量水平

成分	每 100g 含量（鲜重）
水分	94.5 g
能量	18 kcal*
蛋白质	0.88 g
脂肪	0.2 g

(续)

成分	每100g含量（鲜重）
纤维	1.2 g
碳水化合物	2.63 g
酸	0.65 g
钙	10 mg
镁	11 mg
磷	24 mg
钾	237 mg
钠	5 mg
维生素C	14 mg
胆碱	6.7 mg
维生素A和胡萝卜素	0.59 mg
番茄红素	2.57 mg
叶黄素和玉米素	123 g
维生素K	8 mg

* 1 cal=4.186 8 J。

番茄还含有丰富的维生素C。人们为了培育出高类胡萝卜素含量或高维生素C含量的番茄品种做了很多努力，但由于番茄维生素C含量与产量性状存在负相关关系，因此某种程度上目前还没有达到理想的效果（Klee，2010）。

除了这些众所周知的维生素和抗氧化物质，番茄果实中还含有绿原酸、芦丁、质体醌、生育酚和叶黄素等抗氧化物质。此外，还含有膳食碳水化合物、纤维素、风味物质、矿物质、蛋白质、脂肪和生物碱（Davies et al.，1981）。全代谢组分析表明番茄果实中有关初生代谢和次生代谢物质的组分以及这些物质的含量在不同番茄栽培品种及其近缘野生种之间存在广泛的遗传多样性（Tikunov et al.，2005；Schauer et al.，2006；Rambla et al.，2014；Wells et al.，2013；Tieman et al.，2017；Zhu et al.，2018）。

番茄中具有抗氧化活性或有利于人体健康的微量营养物质的遗传变异丰富（Hanson et al.，2004；Schaueret al.，2005），因此，为了提高鲜食和加工番茄的品质，提高果实中这些营养物质的含量特别是类胡萝卜素的含量，长期以来都是主要的育种目标。随着人们对这些营养物质保健功效的认识不断提升，鉴定番茄中影响这些营养物质含量的基因位点已成为新的研究热点。

番茄果实中的维生素类物质主要是维生素 C 和维生素 A。除此之外，番茄果实中还含有叶酸、钾、维生素 E 和其他几种水溶性维生素。截至目前，番茄果实中类胡萝卜素的生物合成过程基本明了，人们发现和鉴定了很多与类胡萝卜素合成相关的基因和突变体（Ronen et al.，1999），其中有 20 多个基因影响番茄果实中类胡萝卜素的种类、数量和分布（Labate et al.，2007）。

Smirnoff 等人（2000）揭示了植物中维生素 C 的代谢途径，Stevens 等人（2007）鉴定出了几个调控维生素 C 含量的数量性状位点（QTLs），Gest 等人（2013）发现维生素 C 含量受品种特性和生长环境的影响。Almeida 等人（2011）鉴定了叶酸合成途径的相关基因，其中有一个调控叶酸含量变化的主效 QTLs 与表观变异相关（Quadrana et al.，2014）。

茄碱及其毒性在茄科作物中普遍存在。番茄中含有 α-番茄碱和脱氢番茄碱，其毒性低于马铃薯中的茄碱（Madhavi et al.，1998；Milner et al.，2011）。Cárdenas 等人（2016）和 Zhu 等人（2018）先后鉴定出了几个调控番茄茄碱含量变化的基因。

番茄果实中矿物质组成主要受植株营养状况的影响，关于矿质元素缺乏的表征及其对植株健康的影响都有详细研究。番茄果实中不同矿质元素含量有很大差异，氮、磷、钾为番茄果实中主要的矿物质，三者含量占果实总无机物含量的 93%（Davies et al.，1981）。基因型对番茄果实矿物质含量有显著影响，不同基因型番茄果实中矿物质含量差异显著。

类黄酮是植物次生代谢物质中一个大的组分，包括花青素、黄酮醇、黄酮、茶多酚和黄烷酮（Harborne，1994）。已经有学者通过转基因来提高番茄果实中类黄酮物质的含量（Muir et al.，2001；Bovy et al.，2002；Colliver et al.，2002）。Willits 等人（2005）在一个野生种中鉴定出了花青素生物合成代谢中的结构基因，该基因在番茄果皮和果肉中组成型表达，潘那利番茄（*S. pennellii*）渐渗系后代的果皮和果肉中均积累了高水平的 5-羟基黄酮。Adato 等人（2009）和 Ballester 等人（2016）在粉果番茄中鉴定出了黄色类黄酮的缺失突变体。

酚酸类也是一个大的组别，茄科植物中咖啡酸前体羟基肉桂酸酯以及绿原酸含量最高（Molgaard et al.，1988）。Rousseaux 等人（2005）注意到果实中抗氧化物质与环境存在很强的互作关系，并在潘那利番茄渐渗系果实中鉴定出了几个调控酚类物质含量的 QTLs。

1.2.2 番茄果实感官品质

长期以来，鲜食番茄的育种目标以提高产量、抗病性以及环境（温室）适应性为主，而对于果实感官品质改良没有明确的目标，因此导致消费者对番茄果实的口感有很多抱怨（Bruhn et al., 1991）。番茄果实的感官品质包括果实大小、颜色、硬度等外在品质和风味、香味、质地等内在品质。番茄果实感官品质受很多因素影响，因此其改良也比较复杂。

番茄的风味品质主要受糖和酸的含量（Stevens et al., 1977）、糖酸比（Stevens et al., 1979；Bucheli et al., 1999）以及挥发性物质组成（Klee et al., 2013）的影响。甜度和酸度与糖和酸的含量及种类有关（Janse et al., 1995；Malundo et al., 1995）。果糖含量对果实甜度的影响大于葡萄糖。果实酸度主要受柠檬酸影响，成熟果实中的柠檬酸含量高于苹果酸（Stevens et al., 1977）。在不同的研究中，果实的酸度主要用果实 pH 或可滴定酸来衡量（Baldwin et al., 1998；Auerswald et al., 1999）。糖和酸对果实整体风味的影响较大（Baldwin et al., 1998）。番茄果实中已经鉴定出的挥发性物质有 400 种以上（Petró-Turza, 1986），仅有一小部分对番茄果实特殊的芳香风味有作用（Baldwin et al., 2000；Tieman et al., 2017）。Verkerke 等人（1998）研究发现，果实的柔韧性与果皮的质地参数相关，因为食用时口腔感觉到的果实硬度与用硬度计测量的果实硬度不完全一致。所以，果实质地性状仍然很难用物理原理测量或是用果实的组成来衡量（Causse et al., 2002），究竟用哪种测量指标最为准确一直是研究者试图要搞清楚的问题（Causse et al., 2010）。

高品质番茄果实的生产受光等环境因素和栽培措施的影响。Davies 等人（1981）、Stevens（1986）和 Dorais 等人（2001）综合论述了可用于番茄品质育种的遗传资源。Causse 等人（2003）研究表明，番茄果实风味和质地性状对消费者的满意度具有重要影响。樱桃番茄的浓厚风味（Hobson et al., 1989）主要源于果实中富含的糖和酸，货架期长的品种比传统品种风味淡（Jones, 1986），这主要是其挥发性物质含量低的缘故（Baldwin et al., 1991）。风味品质作为一种主观感受很难有一个统一标准（Causse et al., 2010）。

番茄的野生近缘种是番茄果实成分改良研究的可用资源。研究人员在克梅留斯基番茄（*S. chmielewskii*）和多毛番茄（*S. habrochaites*）中发现了参与碳代谢酶的突变体，可以改变番茄果实中的糖分组成。克梅留斯基番茄中转化

酶基因突变体 *sucr* 可将果实中的葡萄糖和果糖转化为蔗糖（Chetelat et al., 1995）。多毛番茄中 ADP-葡萄糖焦磷酸化酶的一个等位基因，其基因活性远远高出番茄栽培种中该基因的活性，利用该基因的这一特性可提高栽培番茄果实中的糖含量（Schaffer et al., 2000）；多毛番茄的等位基因 *Fgr* 的其中一个位点可提高果实中果糖和葡萄糖的比例（Levin et al., 2000），该基因是 *SWEET* 基因家族的一员，属于糖转运因子（Shammai et al., 2018）。*Lin5* 基因编码质外体转化酶，是一个调控糖分配的 QTL，潘那利番茄中的等位基因使其果实中糖浓度高于栽培番茄（Fridman et al., 2000）。番茄野生种具有番茄的原始风味，这些风味构成了番茄品质中受喜好的（Kamal et al., 2001）或是不受喜好的味道（Tadmor et al., 2002）。目前，研究人员已经克隆出几个对番茄风味有影响的基因（Klee, 2010；Bauchet et al., 2017a；Bauchet et al., 2017b；Zhu et al., 2019）。

　　果实品质构成因子之间及其与产量、单果重以及果实组分之间的关系复杂，因此，改进果实品质的很多尝试都没有成功。单果重和糖含量通常呈负相关（Causse et al., 2001），但在有些材料中可能成正相关（Grandillo et al., 1996a）。在感官评价和果实组分分析试验中，甜度与还原性糖含量、酸度与可滴定酸含量成正相关（Baldwin et al., 1998；Causse et al., 2002），硬度与仪器测定的硬度成正相关（Lee et al., 1999；Causse et al., 2002），果实大小与抗氧化物的组成相关（Hanson et al., 2004）。高通量代谢组分析可以对果实发育过程或不同基因型之间代谢变化进行全面解析（Schauer et al., 2005；Overy et al., 2004；Baxter et al., 2007）。

　　应生产者和鲜食番茄批发商对果实外观品质的要求，育种家通过利用番茄几个成熟突变体以及影响果实硬度基因的累加效应，使得番茄果实外观和货架期有了很大提升。番茄果实成熟突变体包括目前应用最广泛的成熟抑制基因 *rin*（ripening inhibitor）、不成熟基因 *nor*（non-ripening）和 *alc*（*alcobaca*）。20 世纪 90 年代，长货架期品种番茄就已经进入市场。尽管消费者对长货架期品种番茄的口感有些不满（Jones, 1986；McGlasson et al., 1987），但货架期相关基因的研究还是受到广泛重视，研究者们相继鉴定出了一些影响番茄果实货架期的基因（Vrebalov et al., 2002；Ito et al., 2017；Wang et al., 2019），果实成熟过程中参与细胞壁变化的酶及其对果实硬度和货架期的影响也得到系统研究。结果表明，调控多聚半乳糖醛酸酶和果胶甲酯酶的活性可以改善番茄果实的质地，延长果实货架期（Hobson et al., 1993）。

　　有限生长型番茄受自封顶突变基因 *sp* 的调控。含 *sp* 基因的番茄品种具有开花集中、果实硬度高、抗果实过熟（成熟果实不易过熟）等特性，有利于番茄进行机械化采收。1998 年，Pnueli 等克隆了 sp 基因（Pnueli et al.，1998）。sp 基因是一个基因家族，这个家族至少有 6 个基因成员（Carmel-Goren et al.，2003）。sp 基因不仅影响植株的构造，而且可以调控果重和果实组分相关性状的表达（Stevens，1986；Fridman et al.，2002；Quinet et al.，2011）。最近有研究发现，sp 基因还可以引起番茄植株丧失开花期间对日照长度的敏感性（Soyk et al.，2017；Soyk，2017b）。无离层突变体（j 和 j2）在加工番茄品种中也得到广泛应用，j2 突变体是在克梅留斯基番茄中发现的，其主要作用是使果柄不能形成离层。果柄没有离层，机械化收获时果实就不会携带果柄和萼片（Mao et al.，2000；Budiman et al.，2004），便于后续加工处理。

1.2.3 适度胁迫对番茄果实品质的影响

番茄果实的感官品质受环境和季节的影响。从番茄自身生长发育（周期）来看，光照强度、空气和土壤温度、单株坐果数、植株矿物营养、可利用水分等因素最终都会对果实品质造成影响（Davies et al.，1981；Poiroux-Gonord et al.，2010）。在番茄果实成熟的过程中，温度和辐射通量会影响果实中胡萝卜素、维生素 C 和酚类物质的含量，但对糖和酸含量的影响不大（Venter et al.，1977；Rosales et al.，2007；Gautier et al.，2008）。栽培上，可通过整枝打杈调整单株坐果数。单株坐果数不同，分配到每个果实的碳通量就不同，最终果实的鲜重和干物质含量就不同（Bertin et al.，2000；Guichard et al.，2005）。限制水分供给和用含盐分的水进行灌溉可增加果实中糖的含量（既有浓度效应又有累加效应），提升果实品质，但这种措施对果实中次生代谢物质的含量有反作用（Mitchell et al.，1991；De Pascale et al.，2001；Nuruddin et al.，2003；Johnstone et al.，2005；Gautier et al.，2008；Ripoll et al.，2016），同时，这种措施还有造成减产的风险，减产的幅度取决于处理的浓度、时间长度以及处理时植株所处的发育时期（Ripoll et al.，2014；Guichard et al.，2001；Albacete et al.，2015；Osorio et al.，2014）。

因此，优化栽培技术特别是水分管理是园艺作物生产中果实品质管理的有效措施，同时能兼顾环境因素和消费者需求，只是在应用过程中应尽量将产量损失降到最低（Stikic et al.，2003；Fereres et al.，2006；Costa et al.，2007）。Labate 等人（2007）综合论述了茄科植物对各种气候和营养条件的反应表型，这些表型无论在种内还是种间都表现出广泛的变异性和多样性。番茄对水分和其他非生物因素及其协同限制反应的遗传变异还需要进一步研究，以便挖掘适应这些管理措施的基因型（Poiroux-Gonord et al.，2010；Ripoll et al.，2014）。

为了搞清楚基因型与环境互作（G×E）对番茄果实品质的影响，Auerswald 等人（1999）、Johansson 等人（1999）和 Causse 等人（2003）先后将相同的基因型材料安排在不同地区或不同生长条件下进行重复试验，Semel 等人（2007）、Gur 等人（2011）和 Albert 等人（2016a）开展了水分胁迫和盐胁迫（Monforte et al.，1996，1997a，1997b）对不同基因型材料的品质影响试验。在不同的试验中，基因型与环境互作对番茄果实品质性状，包括果实鲜重、初生和次生代谢物质含量，以及果实硬度均具有显著作用，但是互作效应的贡献比例相对于基因型主效应较小。Albert 等人（2016a）进一步研究基因型与水分互作，将番茄（*S. lycopersicum*）种内杂交得到的重组自交系在两个地区采用两种完全不同的水分管理制度进行试验，结果在两种水分管理制度下有大量的遗传变异，故认为在适度水分胁迫下选育出更高品质的番茄是可行的。

1.3 番茄对生物胁迫的抗性及对非生物胁迫的耐受性

1.3.1 番茄对生物胁迫的抗性

1.3.1.1 番茄病虫害的种类

病虫对露地和设施番茄均有很大的危害。可以侵染番茄的病害至少有 200 种，主要包括细菌、真菌、卵菌、病毒、线虫等引起的病害（Foolad et al.，2012）。害虫主要有蚜虫、蓟马、粉虱、潜叶蝇、食心虫、螨类、叶蝉等，它们会扰乱叶片光合作用中碳同化过程，损坏果实外观，最终导致减产；有些害虫还是病毒传播的载体，有一些病毒如烟草花叶病毒可以通过接触传播。Foolad 等人（2012）汇编了番茄的重要病害，包括 21 种真菌、1 种卵菌、7 种细菌、7 种病毒和 4 种线虫引起的病害。

世界上的番茄产量损失将近 40% 是由病害所致。番茄病害的发生率受地理位置、环境条件以及管理措施等影响。例如，高湿环境有利于链格孢属（*Alternaria*）病菌繁殖，从而引起植物茎部溃疡和早疫病的发生，温暖潮湿的环境有助于匍柄霉属（*Stemphylium*）病菌繁殖，从而引起灰叶斑病的发生，土壤温度过低有利于番茄棘壳孢（*Pyrenochaeta lycopersici*）发生，并可引起植物根腐病发生，气温冷凉有助于镰刀菌属（*Fusarium*）病菌菌落形成，也可诱发根腐病，而空气湿度高且夜温冷凉的环境有利于致病疫霉（*Phytophthora infestans*）生长，从而诱发晚疫病。

1.3.1.2 气候变化对番茄病虫抗性的影响

气候模型预测分析表明，天气的剧烈变化将会导致频繁的干旱和洪涝，使全球温度升高，使农业可利用的淡水资源减少。对植物来讲，面对这些变化，最大的挑战就是要不断提升对各种生物和非生物协同胁迫的抗性和耐受性。无论在露地还是设施中生长的番茄均受到多种非生物胁迫，这些胁迫作用可能增强或是削弱植株对生物胁迫的耐受性。有研究表明，植物对两种及以上因子协同胁迫的响应与对单个胁迫的响应不同，且并不是单个胁迫响应的简单叠加。

长期以来，学界普遍认为 30 ℃ 以上的高温会导致主效抗性基因功能丧失，进而抑制植物防御机制。例如，番茄根结线虫抗性基因 *Mi-1.2* 和番茄叶霉病抗性基因 *Cf-4/Cf-9* 都会在高温下失活（De Jong et al.，2002；Marques De Carvalho et al.，2015）。番茄非生物胁迫与生物胁迫之间存在相互影响关系，如干旱可以降低灰霉病的严重程度，影响白粉病的发生与蔓延。用盐水灌溉番茄会增加尖孢镰刀菌（*Fusarium oxysporum*）和辣椒疫霉（*Phytophthora capsici*）引发病害的可能性，减少新番茄粉孢菌（*Oidium neolycopersici*）的侵染，但不影响灰霉病菌（*Botrytis cinerea*）的侵染（Achuo et al.，2006；Dileo et al.，2010）。Bai 等人（2018）指出，盐胁迫可能通过调节参与信号传导途径的激素平衡来降低 *Ol-1* 基因的抗性水平，但是对 *Ol-2* 和 *Ol-4* 基因的抗性没有影响，这三个基因对番茄白粉病的致病菌新番茄粉孢菌的侵染均有调控作用。限制水分或氮肥供给可增强番茄茎对灰霉病菌的敏感性（Lecompte et al.，2017）。臭氧浓度升高可以引发非常高的环境压力，这种压力可以消除马铃薯纺锤形块茎类病毒（*Potato spindle tuber viroid*，PSTVd）对番茄生物产量的降低效应（Abraitiene et al.，2013）。这里列举的几个例子都是关于环境变化对主要抗性基因所调控的番茄免疫力的影响研究，而有关抗性 QTLs 的研究报道目前还十分有限。尽管基因型与环境互作对生物胁迫抗性影响的研究在逐渐增多，但实际上，人们对植物中参与响应生物和非生物协同胁迫的 QTLs 的研究仍然十分有限。

1.3.1.3 番茄对新病虫害的抗性

全球气候变化会不断引发新的病虫害，使番茄的健康生产受到挑战。一方面，新的病虫种类或株系能够克服既有品种携带的抗性基因；另一方面，世界范围内的农业市场交流也会促使一些新的病虫害的传播，加上气候变化的因素，一些老病害的再次爆发和新病虫害的不断发生，如由致病疫霉引起的晚疫病（Fry et al.，1997），潜叶蝇以及不断增加的新病毒病等。2000 年左右，凤果花叶病毒（*potexvirus pepino mosaic virus*，PepMV）出现，该病毒主要靠机械传播，目前对世界温室番茄生产造成很大威胁（Hanssen et al.，2010）。近期，一种新的烟草花叶病毒——番茄褐皱果病毒（*tomato brown rugose fruit virus*，ToBRFV）在约旦和以色列出现，这种病毒打破了番茄中 *Tm-2* 基因调控的抗性，该基因抗性在过去的 55 年间一直发挥着作用（Maayan et al.，2018）。新病毒的出现常伴随着其介体昆虫的大量繁殖，昆虫和螨类已经对热带地区国家番茄生产造成严重威胁，特别是烟粉虱（*Bemisia tabaci*）可以传播双生病毒，包括已知的番茄黄化曲叶病毒（*tomato yellow leaf curl virus*，TYLCV）和很多其他新生的双生病毒，以及由食心虫引起的作物繁殖阶段的严重问题。利用抗性基因来对抗病毒是防控病毒传播的有效手段之一，但近年来不断出现的新病毒株系打破了已有基因的抗性，从而使抗性基因的功能丧失，如抗性基因 *Sw-5* 对西花蓟马（*Frankliniella occidentalis*）传播的番茄斑萎病毒（*tomato spotted wilt virus*，TSWV）有抗性，也对由正番茄斑萎病毒属病毒（*orthotospovirus*）引起的花生环斑病毒病和最近在美国和加勒比海地区发生的番茄褪绿斑点病毒（*tomato chlorotic spot virus*，TCSV）有抗性，目前这些抗性已经被新的 TSWV 株系克服（Oliver et al.，2016；Turina et al.，2016）。

此外，由番茄细菌性溃疡病菌（*Clavibacter michiganensis* subsp. *michiganensis*，Cmm）引起的番茄细菌性溃疡病肆虐，已对世界番茄生产构成极大威胁。这种病菌由种子传播，是少数经种子传播的病菌种类之一，整个生长过程的植株都要做好防范措施，以防被病菌侵染。从事番茄育种和种苗生产的公司通过良种和植物管理系统（good seed and plant practices，GSPP）有效地保障了番茄种子、植株及种植区域不受 Cmm 浸染。获得 GSPP 认证的区域和公司有权使用 GSPP 商标进行番茄种子和幼苗交易。从 2011 年 7 月开始，GSPP 认证的种子和幼苗已经在法国和荷兰面市。

持续有效地控制这些新的病虫害是全球共同的目标。截至目前，还没有可持续的方法或有效的遗传资源可用于育种以对抗这些新病虫害，可能需要多种遗传策略与栽培管理措施相结合才能有效应对。

1.3.2 番茄对非生物胁迫的耐受性

长期以来，对番茄的驯化和改良主要集中在与产量、品质以及抗病性相关的农艺性状。当前，面对全球气候变化，提升作物的适应性已成为植物育种中一个最具挑战性的课题。增强培育智能型作物品种的意识，鉴定与非生物胁迫耐受性这一新育种目标相关的性状，揭示植物应对环境复杂变化的遗传架构成为培育新品种的关键。实际上，环境因素的变化通常会引起作物在分子水平、生理水平以及形态方面的异常，最终导致农艺性状表型变化。植物逆境适应性在分子水平上要求激活多个逆境响应基因，这些基因参与植物生长中不同的代谢途径，而且其表达受不同转录因子（transcription factors，TF）的调控。基因组学技术加速了不同种间逆境响应基因的鉴定，这些基因属于一个多元化的转录因子家族。其中起主要作用的转录因子家族包括碱性亮氨酸拉链（bZIP）、脱水响应因子结合蛋白（DREB）、APETALA2、乙烯应答元件结合蛋白（AP2/ERF）、锌指蛋白（ZFs）、碱性螺旋-环-螺旋蛋白（bHLH）、热激蛋白（Hsp）等（Lindemose et al.，2013）。这些转录因子在植物中普遍存在并发挥功能，但其功能在不同植物之间具有特异性。

Bai 等人（2018）研究了番茄中已发现的 83 个 *WRKY* 家族基因的功能，发现这些基因在病原侵染、干旱、盐、热及冷胁迫响应中的作用各不相同。*SlWRKY3* 和 *SlWRKY33* 能在干旱和盐胁迫下高表达，可作为候选基因开展进一步研究。对番茄中其他家族基因，包括 *ERFs* 家族基因（Klay et al.，2018）和 *HsP20* 家族基因（Yu et al.，2017）在逆境胁迫中的响应表达也有所研究。除以上家族基因外，单基因也参与番茄非生物胁迫抗性的表达，如 *SlJUB1* 启动干旱抗性，*DREB1A* 和 *VP1.1* 在盐胁迫抗性中起作用，*ShDHN*、*MYB49* 和 *SlWRKY39* 对番茄多种胁迫抗性有作用（Liu et al.，2015；Sun et al.，2015；Cui et al.，2018）。

因为番茄栽培的环境条件宽泛，从露地到温室均有栽培，所以要求番茄具有较强的环境适应能力。同时，番茄是一种用于研究植物对环境响应的遗传、揭示基因型与环境互作机制的理想模式植物。

1.3.2.1 水分胁迫

番茄是一种需水量较大的作物（Heuvelink，2005），水分管理是番茄生产管理的关键。番茄生产中适宜的灌水量取决于植株蒸腾作用和发育阶段。生长过程中缺水会对产量造成影响。在研究水分胁迫对番茄影响的试验中，多数是以最适需水量为基准，用百分比来表示水分胁迫的程度（Albert et al.，2016a，2016b；Ripoll et al.，2016；Diouf et al.，2018）。

从农艺学角度出发，减产是番茄缺水最主要的后果，不同发育阶段番茄对缺水的敏感程度因品种特性和胁迫强度而异，如果缺水发生在果实发育阶段，那么减产会更严重（Chen et al.，2013）。种子萌发是植物感受环境胁迫的第一步，番茄种子萌发阶段渗透胁迫会抑制种子萌发（Bhatt et al.，1987）。在营养生长和生殖生长阶段，水分胁迫对作物的经济表型有负效应，而对果实品质有正效应。Costa 等人（2007）介绍了水分胁迫对果树和包括番茄在内的果菜作物两个方面的影响：一方面造成减产；另一方面会提高果实的品质，如果实的维生素 C、抗氧化物质和可溶性糖的含量会增加。因此，在实际生产中存在一个如何权衡产量和品质的问题（Albert et al.，2016a；Ripoll et al.，2014；Patanè et al.，2010；Zegbe-Domínguez et al.，2003）。大果番茄和樱桃番茄对环境胁迫的敏感性不同，对水分胁迫与对照处理下不同番茄品系的分析表明，大果番茄对水分胁迫更敏感，响应更强烈（Albert et al.，2016b）。试验还发现，水分胁迫可使番茄果实中糖的相对含量增加，其原因是果实中水分含量的降低而不是糖合成的增加。然而，Ripoll 等人（2016）发现，在番茄果实发育的不同阶段进行水分胁迫，有助于果实中果糖和葡萄糖的合成增加。这两个试验说明，在水分胁迫下，稀释效应和糖合成增加共同作用，提高了番茄果实的品质。多组学分析为研究不同环境条件下目标基因的表达水平及其变化提供了有效途径。番茄中参与缺水响应及干旱耐受性的基因已有报道，*SlSHN1* 基因就是一个例子，其通过激活叶片中参与角质层蜡质积累的相关下游基因诱导番茄的抗旱性（Al-Abdallat et al.，2014）。Wang 等人（2018）鉴定出了一个干旱诱导基因 *SlMAPK1*，其主要作用是激活抗氧化酶活性，消除已产生的 ROS，进而使作物对干旱具有较强的耐受性。

1.3.2.2 盐胁迫

土壤盐渍化已经成为全球农业特别是地中海地区农业的一大问题，土壤干旱和非持续灌溉使盐渍化土壤面积不断增加（Munns et al.，2008）。Munns 等人（2015）对盐胁迫（SS）给出了定义：当盐分积累超过植物用于生长的量，植物需要对盐分积累进行防御时，就构成了盐胁迫。番茄可以耐受的盐浓度为 2.5dS/m，樱桃番茄对盐胁迫的耐受性高于大果番茄（Scholberg et al.，1999；Caro et al.，1991）。产量是衡量盐胁迫耐受性的一个指标，盐分浓度超过耐受阈值会引起产量显著下降，盐胁迫下番茄产量的减少与果实大小和数量的减少相关联（Scholberg et al.，1999）。与水分胁迫效应相同，盐胁迫也会使番茄果实中糖含量增加（Mitchell et al.，1991），还可引起果实中正负离子比例的变化。果实中糖含量增加可能是水分含量降低、离子浓度增加以及糖积累几个方面相互作用的结果（Navarro et al.，2005），这些变化都是番茄植株对渗透调节响应的结果。上面提到的番茄盐分耐受阈值是从特定的几个番茄品种的研究中得出的。Alian 等人（2000）注意到，鲜食番茄品种对盐胁迫响应在不同基因型之间的差异很大，换言之，可以通过鉴定筛选出耐盐性强的番茄材料用于品种选育。

植物对盐胁迫的一系列响应过程，就是进行细胞稳定性的再平衡和再建立的过程。植物对盐胁迫的生理反应高度依赖离子通道转运体调节离子平衡（Apse et al.，1999）。Rajasekaran 等人（2000）通过试验筛选出了一些对盐胁迫耐受性强的番茄野生近缘种，发现其盐胁迫耐受性主要与根部对高 K^+/Na^+ 比例的耐受性有关。醋栗番茄（*S. pimpinellifolium*）在盐胁迫反应中，产量和存活相关性状的遗传变异性很高（Rao et al.，2013）。在产量构成性状中，果实数量是最易受到盐胁迫影响的性状，在野生种和栽培种中表现一致（Rao et al.，2013；Diouf et al.，2018）。番茄遗传资源中具有丰富的可利用的变异，因此，通过筛选盐胁迫条件下生理性状或农艺性状表现优良的品种，就有可能培育出耐盐的番茄品种。

1.3.2.3 温度胁迫

所有作物都有一个适宜生长的温度范围。众所周知，番茄对环境温度的适应范围很广，从高海拔低温地区到热带干旱高温地区都可以生长。根据作物模拟生长模型，Boote 等人（2012）指出，番茄生长和果实发育最适温度为 25 ℃左右，温度低于 6 ℃或高于 30 ℃，植株生长、授粉和果实发育都会严重受阻，最终会对果实产量产生负面影响。开展不同番茄材料及其野生近缘种对低温和高温胁迫响应的差别研究，有助于认识和理解作物对温度胁迫的反应机制。

（1）高温胁迫。 全球温度上升是气候最显著的变化趋势。预计到 21 世纪末，全球气候变暖将导致世界范围内主要栽培作物产量显著下降（Zhao et al.，2017）。根据植物受高温胁迫（HT）时间的长短，可将高温胁迫分为短期热胁迫和长期热胁迫。当植物受到的高温胁迫表现为间断、波动和不连续时，即为短期热胁迫；如果植物连续几天经受高温，则为长期热胁迫。长期热胁迫对作物的农艺性状有很大影响，尤其是在作物整个生长季节中发生长期热胁迫。在露地条件下，土壤高温与空气高温的影响作用不同，土壤高温普遍影响种子萌发。种子不同发育时期对高温的敏感程度不同，开花期是作物对高温胁迫最敏感的时期（Wahid et al.，2007），因为其生殖器官受到高温胁迫会导致生殖过程受阻，这也是高温胁迫造成严重减产的主要原因（Nadeem et al.，2018）。高温胁迫对番茄花期生殖生长的抑制作用主要表现为对雄性育性有很大的改变，而对雌性育性的影响较小（Xu et al.，2017a，2017b）。因此，普遍认为雄性育性改变是高温胁迫下影响番茄生殖的主要因素，在有关番茄耐热性的研究中，多采用花粉性状替代唯一的产量指标进行材料耐热性评价

（Driedonks et al.，2018）。与栽培品种相比，番茄野生种雄性生殖性状变异性高。有些材料在高温胁迫下，花粉活力仍然很高，将这些野生种中的耐热等位基因转入栽培番茄中，有望改善栽培番茄对高温的耐受性。有人在培养箱中对栽培番茄进行 26 ℃ 的高温胁迫，结果单性结实率提高，而坐果率降低（Adams et al.，2001）。同时观察到，高温会加速果实的成熟过程，因此也会影响果实最终的品质。

由于高温对农业生产有着重要影响，针对高温胁迫的研究已经鉴定出一些热响应基因（Waters et al.，2017；Keller et al.，2018；Fragkostefanakis et al.，2016）。这些热响应基因通常由热胁迫转录因子（HSF）的活性来进行调控，这些热胁迫转录因子在相关文献中都有具体介绍。研究主要的热胁迫转录因子在不同作物中的耐热性功能有助于通过基因编辑技术培育耐热番茄（Fragkostefanakis et al.，2015）。

（2）低温胁迫。番茄野生种的地理分布区域广泛，其中就包括年均气温低于栽培番茄生长最适宜温度的高海拔地区（Nakazato et al.，2010），这揭示了番茄具有适应亚适宜温度条件的潜能。

Adams 等人（2001）观察到，14 ℃ 时番茄生长缓慢。Ploeg 和 Heuvelink（2005）注意到，当温度低于 12 ℃ 时，几乎观察不到番茄的生长。与高温胁迫类似，番茄在低温胁迫下花粉活力下降，坐果受阻，花数量、果实数量以及产量均随温度降低而降低。低温严重影响番茄的光合作用，也影响光合产物运输和分配的效率（Meena et al.，2018）。相对水分含量、叶绿素荧光以及酚类化合物含量等与耐冷机制有关的指标，常用来表示植物对低温胁迫的耐受性（Giroux et al.，1992；Dong et al.，2019；Khan et al.，2015）。Meena 等人（2018）发现，外源施用酚类物质如水杨酸可以显著提高番茄对低温的耐受性。低温胁迫对番茄的果实品质有很大影响。番茄植株如果在生长发育过程中受到低温胁迫，那么果实中非酶类抗氧化物如番茄红素、β-胡萝卜素和 α-生育酚含量就会降低。

转录组分析有助于挖掘与番茄低温胁迫相关的候选基因。Zhuang 等人（2019）通过转录因子分析，鉴定出一个番茄的低温胁迫响应基因 *SlWHY1*，该基因在 4 ℃ 低温时表达增强，对叶绿体中的光系统Ⅱ具有保护作用，能促进叶绿体中的淀粉积累。在有些植物中，低温胁迫信号与 C-重复结合因子（C-repeat binding factor，CBF）有关（Jha et al.，2017），CBF 可激活下游低温胁迫响应基因产生对低温胁迫的抗性，番茄中对低温胁迫适应性具有调控作用的主要 CBF 都有相关研究报道（Mboup et al.，2012）。Kenchanmane Raju 等人（2018）指出，植株在响应低温时，与光合作用和叶绿体发育相关基因的表达均会受到持续抑制，相反，表现上调且最保守的基因大部分都属于 *CBFs*、*WRKYs* 和 *AP2/EREBP* 转录因子家族基因。以上结果表明，这些转录因子家族基因可用于耐低温番茄品种的选育。

1.3.2.4 矿质营养胁迫

众所周知，矿物质对植物生长具有促进作用。通常矿质元素分为必需元素和非必需元素两种。非必需元素对植物发育大有裨益（Marschner，1983），而必需元素是植物生长的基础营养元素。氮（N）、磷（P）、钾（K）是高等植物最重要的必需元素，它们的使用具有显著的环境成本，因此，筛选低需肥作物有助于培育智能型作物品种。

（1）氮胁迫。氮是番茄发育中最重要的限制性营养元素，缺氮会对番茄重要经济性状造成严重影响。研究表明，缺氮对番茄果实数量、大小、储藏品质、颜色以及口感均产生负效应（Sainju et al.，2003）。de Groot 等人（2004）和 Larbat 等人（2012）的数据显示，番茄生长速率与氮供应量存在线性相关关系。低氮限制番茄叶部生长，却能促进其根系发育，这种作用主要与细胞分裂素浓度变化有关。Larbat 等人（2012）发现，氮素不足会引起酚类物质合成途径相关基因表达上调，因此，酚类化合物积累量增加是番茄氮供应不足的一个显著标志。

过量施用氮肥除了会对地下水造成污染外，也会使植株营养生长过旺，并影响果实发育，加速果实成熟，限制植株根系发育（Du et al.，2018）。因此，要实现对番茄生产过程中氮的科学管理，一方面，要从遗传改良入手，研究与氮利用效率相关的基因，并对这些基因加以正确利用，提升番茄对氮缺乏的耐受性；另一方面，要采用先进的温室管理技术如计算机智能系统，以及类似叶片发射率等新的胁迫标记性状来实现氮素的合理化管理（Elvanidi et al.，2018）。

（2）磷胁迫。土壤中的磷通常不能被植物直接利用，因此施肥是向作物（包括番茄）提供磷营养的必要措施。植物从土壤中获取磷的能力与根系形态以及植物激素水平变化有关，植物发育早期对磷非常敏感，磷供给不足会导致植株生长和发育不良（Sainju et al.，2003；De Groot et al.，2004），磷不足会引起根系形态结构变化，可以通过增强生长素的敏感性激活磷转运基因，将脂类和核酸中的磷变为可移动和可利用的状态（Schachtman et al.，2007）。长期磷饥饿会抑制初生根的生长而促进次生根的生长（Xu et al.，2012），使叶片净光合速率降低，蔗糖含量降低，淀粉含量增加。上述研究人员已从番茄中鉴定出了不同的磷饥饿响应基因,这些基因均属于编码磷酸丝氨酸结合蛋白的 14-3-3 基因家族,编码的磷酸丝氨酸结合蛋白主要参与蛋白之间的互作。

大田条件下,土壤中有足够的、可吸收利用的磷是根系良好发育的保障;温室条件下,磷的施用是可控的。要实现对磷的合理施用,提升番茄对磷缺乏的耐受性,就需要对磷缺乏响应基因有更多研究和了解,掌握磷缺乏相关基因及其与形态和生理反应的关系,有利于培育磷利用效率高的番茄品种。

(3) 钾胁迫。钾元素在植物营养中的重要性主要体现在参与重要的生理过程,如光合作用、渗透调节作用和离子稳定调节作用(Marschner,1983;Pettigrew,2008)。植物产量和品质受光合能力影响,而光合能力又与植物器官中钾离子浓度直接相关。番茄上施钾的正效应表现为:植株生长旺盛、提早开花、果实数量增加、可滴定酸比率增加(Sainju et al.,2003)、果实中可溶性固形物含量增加、抗氧化能力增强、维生素 C 含量增加(Tavallali et al.,2018)。相反,缺钾会引起形态伤害,导致番茄叶片黄化、叶脉间褪绿以及叶缘褐化干枯。植物感应外部钾离子浓度变化首先激活信号转导,启动重建新的离子平衡维持离子稳定性。植物对低钾的适应通过对不同钾离子转运体的移动调控来实现,Wang 等人(2015)介绍了参与钾离子转运体移动的通道的功能和作用,包括对钾离子运输起关键作用的 HAK/KUP/KT 转运子家族。植物中钾离子的运输首先从根部开始,因此缺钾对根系结构有很大影响(Zhao et al.,2018),改善根系发育可直接减小缺钾造成的不良影响。

(4) 钙胁迫。钙元素是一种非常重要的离子,广泛参与植物生长发育中关键性的代谢过程(Bush,1995)。钙作为一种常量元素,在植物中已经明确的功能包括多个方面,其中在维持细胞壁刚性和细胞膜稳定性、调控离子转运以及非生物胁迫的信号转导等方面具有重要作用(Hepler,2005;Hirschi,2004;Wilkins et al.,2016)。植物钙缺乏会引起很多问题,钙离子浓度水平与细胞离子稳定性变化相关联,与营养不平衡程度有关。钙离子减少与环境诱导一样,首先会引起胞质中钙离子浓度改变,钙离子浓度改变通过转运体蛋白介导调节钙离子流,以重建离子稳定性(Bush,1995)。此外,植物对非生物胁迫的响应也与钙离子的稳定性密切相关,钙离子的稳定性是植物耐受性信号转导和建立的关键(Rengel,1992;Wilkins et al.,2016)。对番茄的研究表明,钙离子营养可以缓解盐胁迫对植株和果实生长造成的负面影响(Tuna et al.,2007)。钙离子利用率与植物中充足的钙离子有关。钙离子利用率是植物对环境胁迫适应性的一个重要特征,这一性状的遗传变异性表明,培育适应低钙离子摄入的品种具有很大潜力(Li et al.,1990)。然而,大多数番茄都对钙离子缺乏反应敏感，并会出现不良反应，如生理性病害脐腐病（BER）（Adams et al.，1993）。也有研究表明脐腐病与基因型有关，有些基因型能够通过增加抗氧化代谢物如抗坏血酸盐的合成削弱氧化胁迫（Rached et al.，2018）；有些基因型对赤霉素敏感，表明细胞中钙缺乏不是脐腐病的直接诱因。此外，De Freitas 等人（2018）通过转录组分析，鉴定出了抑制番茄脐腐病的候选基因，这些基因主要与氧化胁迫的抗性有关。番茄脐腐病是一个复杂的生理性病害，诱发因素包括非生物胁迫、基因型、生理或农艺因子及它们之间相互作用（Hagassou et al.，2019）。然而，考虑到番茄脐腐病与钙离子浓度水平紧密关联，在不同环境诱导条件下，鉴定出参与调控钙离子稳定性的通道基因家族有助于揭示脐腐病发生与钙离子浓度之间相互作用的分子机制。

1.3.3　番茄对综合胁迫的耐受性

关于番茄的胁迫耐受性研究，以往主要集中在植株某一生长阶段对单一胁迫的响应以及通过何种途径增强植物对胁迫的耐受性。然而，在自然条件下，通常是多个胁迫因子同时作用，尤其随着气候变化，多种环境胁迫如水分胁迫和高温胁迫常常协同发生。气候变化对病菌的扩散、新病害的出现和传播有直接影响（Harvell et al.，2002），生物胁迫和非生物胁迫协同作用呈上升趋势，胁迫组合的形式因地理位置和种植区域而异。Suzuki 等人（2014）根据不同作物受到的不同胁迫建立了一个胁迫矩阵，展示了不同胁迫组合可能产生的正面效应和负面效应。胁迫组合对产量、形态、生理性状的整体影响可能与单个胁迫因子的影响明显不同。因此，可以借鉴此矩阵建立番茄的胁迫矩阵，这将有助于全面了解协同胁迫对番茄植株的影响，进而有针对性地提出番茄育种方案。

关于协同胁迫对番茄不同性状的影响已有研究和文献报道。Zhou 等人（2017）研究表明，在水分胁迫和高温胁迫的组合胁迫下，不同品种的番茄在植株生理和生长性状响应模式上基本相似，而对单独高温胁迫的响应则不同，高温胁迫和盐胁迫组合胁迫对番茄生长的危害作用较单一盐胁迫要小（Rivero et al.，2014）。除了形态变化，在拟南芥上的试验发现，组合胁迫时的基因表达变化与单个因子胁迫时的基因表达变化各自独立并无关联（Rasmussen et al.，2013）。

除了不同环境因子组合胁迫，生物和非生物因子协同胁迫将会普遍发生，特别是在大田条件下，而以往的研究都是将两者分开单独进行。近年来，已经有相关研究填补了这方面认知的空白。Kissoudisetal 等人（2015）用盐和新番茄粉孢菌（*Oidium neolycopersici*）两者同时胁迫番茄，发现盐胁迫增加了重组群体对白粉病的敏感性。Anfoka 等人（2016）发现，长时间高温胁迫伴随着番茄中番茄黄化曲叶病毒（TYLCV）的积累量减少，这是植物对高温胁迫响应作用的结果。胁迫响应如内源激素分泌和 ROS 的产生，都是植物对生物和非生物胁迫响应的重要生理表现（Fujita et al.，2006），这些生理过程需要调控两种胁迫相关基因的共同参与。番茄中有一些基因参与生物和非生物胁迫应激响应，如 *SlGGP-LIKE*，Yang 等人（2017）发现该基因可以促进维生素 C 合成，降低 ROS 的伤害，提高番茄对冷害的耐受性，而抑制该基因表达会引起 ROS 积累，产生并增强对丁香假单胞菌（*Pseudomonas syringae*）的抗性。Ashrafi-Dehkordi 等人（2018）通过转录组分析，在番茄中发现了一组基因，其表达受生物和非生物协同胁迫的影响。番茄中也鉴定出了一些单基因，这些基因参与响应非生物胁迫和丁香假单胞菌（Sun et al.，2015）或疫霉菌（*Phytophthora infestans*）（Cui et al.，2018）的协同胁迫，并可以用于抗性育种。然而，截至目前，对不同组合胁迫的研究大多停留在基因组学和代谢组学分析层面，而从遗传学角度进行的研究仅局限于拟南芥，在其他作物上的研究几乎没有（Thoen et al.，2017）。

矿质元素对植物病原也有重要影响。施氮肥可以增加酚类物质和挥发性物质的积累，这些物质可以诱发番茄植株对由白粉虱、烟粉虱（Islam et al.，2017）和潜叶蝇（Han et al.，2015）侵染而引发的病害防御反应。Lecompte 等人（2010）报道了氮与番茄对灰霉病菌（*Botrytis cinerea*）抗性之间的相互作用关系。在番茄中，施氮不仅对生物胁迫的耐受性有作用，而且对一些非生物胁迫的抗性也有不同程度的影响。

非生物胁迫中，盐胁迫是对番茄胁迫响应影响最大的一种胁迫。Papadopoulos 等人（1983）研究了盐胁迫和施氮的协同效应，发现施氮对生长和果实重量具有正面效应，而这种效应可以被浓度达到 5dS/m 的盐胁迫抑制。

Frary 等人（2011）研究表明，在番茄种间渐渗系群体中，盐胁迫使得叶片中的 Ca^{2+} 和 K^+ 含量分别下降 47% 和 8%；潘那利番茄中等位基因对维持盐胁迫和对照处理中高 Ca^{2+} 含量起主要作用，潘那利番茄可以用来提升番茄对盐胁迫和低 Ca^{2+} 的耐受性。

2 番茄的遗传和基因组资源

2.1 番茄的遗传资源

2.1.1 番茄的起源及其野生种

农业和食物遗传资源是全球营养和食品安全的关键(FAO, 2015)。在作物生产过程中,保持遗传多样性不仅是培育新品种、鉴定目标性状候选基因、解析植物进化历史的重要手段,也是减少生物和非生物胁迫危害等的根本保障。

番茄属于茄科,茄科是一个多样性广泛的大科,包含 3 000 多个种,其中主要作物如茄子起源于亚洲,辣椒、马铃薯、烟草、番茄则起源于南美洲。番茄进化枝(*Lycopersicon*)包含栽培番茄(*Solanum lycopersicum*)及其 12 个野生近缘种(Peralta et al., 2005)。早在 20 世纪 40 年代,Charles Rick 和他的同事就率先开展对番茄野生近缘种的勘探和研究工作。

番茄进化枝中的所有种均起源于安第斯山脉,包括秘鲁、玻利维亚、厄瓜多尔、哥伦比亚和智利,其生存环境从海平面到海拔 3 300 m,从干旱地区到多雨气候地区,从安第斯高地到加科隆群岛沿岸。番茄通常生长在狭窄且相互分隔的山谷,因此能适应不同类型的土壤和多种气候条件。生存环境的广泛性造就了番茄野生种的遗传多样性,也使番茄在形态、生理、性型及分子水平等方面具有广泛的多样性(Peralta et al., 2005)。

番茄的驯化始于几千年前,源于醋栗番茄(*S. pimpinellifolium*),经历了两个阶段。首先在秘鲁,醋栗番茄被驯化形成了樱桃番茄(*S. lycopersicum cerasiforme*);然后在墨西哥,樱桃番茄进化形成了大果番茄(Bauchet et al., 2012),这一进化过程在分子水平上得到了确认(Blanca et al., 2012; Lin et al., 2014; Blanca et al., 2015)。仅有一小部分大果番茄种子从墨西哥被带回欧洲,经不断驯化后遇到了新的遗传改良瓶颈。番茄栽培首先在欧洲南部缓慢扩散,直到二战后才开始被选择性种植,随之遍布全世界。

2.1.2 番茄的种质资源库

当前全世界拥有番茄资源 83 000 份以上,保存于不同的种子库中(FAO,

2015），包括位于美国加利福尼亚州戴维斯市的番茄遗传资源中心（TGRC）、位于纽约州的美国农业部（USDA）、位于中国台湾的世界蔬菜中心（AVRDC），以及荷兰遗传资源中心（CGR）等，这些种子库保存着番茄大多数遗传资源。

加利福尼亚大学戴维斯分校番茄遗传资源中心拥有世界上数量最多的番茄野生近缘种资源，这些资源都是 Charles Rick 在其有生之年勘察搜集的，是番茄育种和新基因挖掘的重要源泉。例如他发现的 46 个潘那利番茄（*S. pennellii*）对干燥环境具有特殊适应性，但这些资源仅在秘鲁发现，其他地区尚无分布（Tomato Genetics Resource Center, University of California, Davis, https://tgrc.ucdavis.edu/Data/Acc/Wildspecies.aspx）。

2.1.3 番茄的突变体库

自然遗传的多样性是作物适应性育种的主要源泉。自然突变可以出现在栽培种中，也可以出现在野生近缘种中。突变是很多性状遗传多样性的源泉，包括抗病基因和品质性状相关基因（Bauchet et al., 2012；Bauchet et al., 2017a；Rothan et al., 2019）。然而，已经被克隆并验证了其具体功能的基因的数量仍然很有限（Rothan et al., 2019），有些生物技术工具如定向诱导基因组局部突变技术（targeting induced local lesions in genomes，TILLING）（Comai et al., 2006）可以创制特定材料的突变体库，加速功能组学研究，发掘在已知位点的有用等位基因（Menda et al., 2004；Baldet et al., 2007；Okabe et al., 2011；Mazzucato et al., 2015；Gauffier et al., 2016）。这项技术最典型的应用，就是通过化学诱变剂使基因组发生碱基突变。世界上已有几个番茄 TILLING 突变体库，如美国加利福尼亚大学的 UCD 基因组中心 TILLING 实验室、日本的 Microtom 突变体库（Okabe et al., 2011）、TOMATOMA 数据库、印度海德拉巴大学番茄基因组资源库、法国农业科学院番茄 TILLING 平台（Minoïa et al., 2010）、意大利麦塔庞顿农业生物学 LycoTILL 数据库等。

番茄及其近缘野生种请参考表 2-1。

表 2-1 番茄及其近缘野生种（Peralta et al., 2005）

种	分布	原生地（海拔范围）	类别
Solanum lycopersicum L.	全球种植，栽培种	栽培，海平面至海拔 4 000 m	Lycopersicon group
Solanum pimpinellifolium L.	厄瓜多尔西南部到智利北部（厄瓜多尔北部多与 *Solanum lycoper-sicum* 混生）（Peralta et al., 2005；Blanca et al., 2013)	干燥坡地、平原及耕种田周边，海平面至海拔 3 000 m	Lycopersicon group

2 番茄的遗传和基因组资源

(续)

种	分布	原生地（海拔范围）	类别
Solanum peruvianum L.	秘鲁中部到智利北部	干燥海岸沙漠，海平面至海拔 3 000 m	Eriopersicon group
Solanum cheesmaniae (L. Riley) Fosberg	科隆群岛	干燥、开阔、多石坡地，海平面至海拔 1 300 m	Lycopersicon group
Solanum galapagense S. C. Darwin and Peralta	科隆群岛	干燥、开阔、多石坡地，海平面至海拔 1 600 m	Lycopersicon group
Solanum arcanum Peralta	秘鲁北部	干燥安第斯山谷和海岸季节性多雾潮湿生境，海拔 100~4 000 m	Arcanum group
Solanum chmielewskii (C. M. Rick，Kesicki，Fobles & M. Holle) D. M. Spooner，G. J. Anderson & R. K. Jansen	秘鲁南部和玻利维亚北部	干燥安第斯山谷，多见于开阔、多石坡地及路沿，海拔 1 200~1 300 m	Arcanum group
Solanum neorickii D. M. Spooner，G. J. Anderson & R. K. Jansen	厄瓜多尔南部到秘鲁南部	干燥安第斯山谷，海拔 500~3 500 m	Arcanum group
Solanum chilense (Dunal) Reiche	智利海岸和秘鲁南部	干燥、开阔、多石坡地，海平面至 4 000 m	Eriopersicon group
Solanum corneliomulleri J. F. Macbr.	秘鲁南部	干燥、多石坡地，海拔 20~4 500 m（低海拔群落多在秘鲁南部山坡或悬崖崩塌处）	Eriopersicon group
Solanum habrochaites S. Knapp and D. M. Spooner	秘鲁和厄瓜多尔的安第斯山脉	山地森林，干燥坡地及海岸，海拔 10~4 100 m	Eriopersicon group
Solanum huaylasense Peralta	圣克鲁斯河流域和秘鲁中北部	干燥、开阔、多石坡地，海拔 950~3 300 m	Eriopersicon group
Solanum pennellii Correll	秘鲁北部到智利北部	干燥坡地和冲刷地，通常平坦地域海平面至 4 100 m 海拔	Neolycopersicon group

注：Lycopersicon group 对应红色和橙色果实种，关于野生种的杂交亲和性及其他生物学参数请参考 Grandillo 等人于 2011 发表的文章。

2.2 番茄的分子标记、基因和 QTL

2.2.1 番茄分子标记的发展

在分子标记发现之前，番茄作为模式植物已经广泛用于对目标性状的遗传研究和突变基因作图定位（Butler，1952）。番茄中首个目标基因定位归功于近等基因系（NIL），因为成对近等基因系之间仅在目标基因区域有差异，而其他背景相同（Philouze，1991；Laterrot，1996），然而，直到 20 世纪 80 年代，突变基因在遗传图谱上的定位仍然不够准确。同工酶标记由于数量有限，很快被限制性内切酶片段长度多态性（RFLP）分子标记所取代。Tanksley 等人构建了第一个基于 RFLP 分子标记的高密度遗传连锁图谱（Tanksley et al.，1992），12 条染色体共有 1 000 多个位点，给出了人们感兴趣的突变和基因的位置。接着，以 PCR 为基础的分子标记，包括随机扩增多态性 DNA（random amplified polymorphic DNA，RAPD）、扩增片段长度多态性（amplified fragment length polymorphism，AFLP）及微卫星等逐步被发现和应用，但是这些标记在多态性和基因组区域覆盖方面仍然有局限性。随着目标基因相关分子标记的发现，特异 PCR 标记建立，这简化了育种中对基因型选择的过程。然而，由于大多数 PCR 标记如 RAPD 和 AFLP 标记靠近着丝点，因此降低了番茄基因定位的效率（Grandillo et al.，1996a；Haanstra et al.，1999；Saliba-Colombani et al.，2001）。

2.2.2 番茄数量性状 QTL

利用分子标记构建的遗传图谱将数量性状解析为数量性状位点（QTL）（Paterson et al.，1988；Tanksley et al.，1992），从此开辟了数量性状遗传因子物理图谱构建及其分子克隆的新途径（Paterson et al.，1991）。番茄上图位克隆的第一个基因是番茄细菌性斑点病抗性基因 *Pto*（Martin et al.，1994）。随后，对每个野生近缘种和栽培种种间杂交后代的研究陆续展开。由于栽培种的遗传多样性匮乏（Miller et al.，1990），大多数作图群体都是由栽培种和来自番茄进化枝中的野生近缘种种间杂交（Foolad，2007；Grandillo et al.，2011），或与类番茄茄（S. *lycopersicoides*）（Pertuzé et al.，2003）和龙葵组（Albrecht et al.，2010）杂交后代构建而成。而基于种内杂交后代群体构建的图谱多用于番茄果实品质性状的解析。研究人员利用这些种间或种内杂交后代构建的群体发现并鉴定出了很多主效基因和不同性状的 QTLs（Grandillo et al., 1996b; Tanksley et al., 1996; Fulton et al., 1997; Bernacchi et al., 1998a, 1998b; Chen et al., 1999; Grandillo et al., 1999; Frary et al., 2000; Monforte et al., 2000; Causse et al., 2001; Saliba-Colombani et al., 2001; Causse et al., 2002; Doganlar et al., 2003; Frary et al., 2004; Schauer et al., 2006; Baldet et al., 2007; Jiménez-Gómez et al., 2007; Cagas et al., 2008; Kazmi et al., 2012a, 2012b; Haggard et al., 2013; Alseekh et al., 2015; Pascual et al., 2015; Ballester et al., 2016; Rambla et al., 2014; Kimbara et al., 2018)。

番茄 QTLs 的主要研究进展可归纳如下。

①发现的 QTLs 较多，但具有强效应的 QTLs 较少。几乎每个研究都能够找到相应的 QTLs，然而，大多数情况下所获得的能够解释大部分表型变异的 QTLs，通常都需要与微效 QTLs 协同作用，仅在少数情况下才能找到具有强效应的 QTLs。目前已经鉴定出的 QTLs 大多数表现为加性效应，仅有一小部分表现为显性或是超显性效应（Paterson et al., 1988; De Vicente et al., 1993）。

②QTLs 可分为两种类型：一种属于稳定型，不受环境、年代和后代群体的影响；另一种则具有较强的专一性，仅在某种条件下发挥效应（Paterson et al., 1991）。

③参与某个性状变异的区间 QTLs，可以从同种不同品系构建的群体中获得，也可以从不同种间构建的群体中获得（Fulton et al., 1997; Bernacchi et al., 1998a, 1998b; Chen et al., 1999; Grandillo et al., 1999; Fulton, 2002）。

④将复杂性状解析为相关构成性状并对构成性状进行 QTL 作图，有助于认识和理解这些复杂性状变异的遗传基础。例如，调控鲜食番茄感官品质的几个构成因子的 QTLs 图谱，揭示了感官因子 QTLs 与果实化学组成之间的关系（Causse et al., 2002）。因此，分析一个性状的生物化学组成也尤为重要。

⑤精细定位可以准确给出 QTLs 在染色体上的位置，证明同一染色体区域中几个 QTLs 之间的连锁关系（Paterson et al., 1990; Frary et al., 2003; Lecomte et al., 2004a）。Eshed 等人（1995）在同一个染色体臂上鉴定出了 3 个控制果实重量的连锁 QTLs。精细定位也是 QTLs 克隆的重要步骤，首个成功的范例就是调控果实重量（Alpert et al., 1996; Frary et al., 2000）、果实形状（Tanksley, 2004）和可溶性固形物含量（Fridman et al., 2000, 2004）的 QTLs 克隆。

⑥与番茄栽培种相比，对番茄野生种的研究还很欠缺，而野生种所携带的等位基因可用于大多数农艺性状的改良和提高（De Vicente et al.，1993）。

2.2.3 番茄表型解析专用群体库

分子育种技术一经形成就迅速被用于挖掘"聚合育种"基因和农艺性状相关的目标QTLs，首个应用范例就是高代回交QTL方法（advancedback-cross QTL，AB-QTL）（Grandillo et al.，1996b）。Grandillo等采用AB-QTL法，从栽培番茄×醋栗番茄（*Solanum lycopersicum* × *Solanum pimpinellifolium*）高代回交后代中，挖掘出了农艺性状有益QTLs的等位基因。这一结果说明，可以通过这种途径利用番茄野生种对番茄栽培种进行改良（Grandillo et al.，1996）。

种间杂交渐渗系（introgression lines，IL）是利用野生资源的又一途径。渐渗系中的供体染色体片段通常来源于野生种，利用种间杂交渐渗系我们可以解析供体染色体片段的功能，也可以对相应QTLs的农艺表现进行评价（Paran et al.，1995），种间杂交渐渗系也是QTL精细定位和图位克隆的基础。第一个种间杂交渐渗系文库是用潘那利番茄和栽培番茄构建的（Eshed et al.，1995；Zamir，2001）。与双等位基因QTL作图群体相比，种间杂交渐渗系的QTL作图效力增强，特别是亚渐渗系（sub-IL）拥有更小的导入片段，使其QTL作图效力更强。利用这类群体已经成功鉴定出了控制番茄果实性状（Causse et al.，2004）、抗氧化物（Rousseaux et al.，2005）、维生素C（Stevens et al.，2007）和挥发性芳香物质的QTL（Tadmor et al.，2002）。育种家们利用这些从渐渗系中鉴定出的QTLs显著提高了商业品种番茄果实中的固形物含量，并大幅度提升了番茄的产量（Fridman et al.，2004）。

回交自交系（backcrossed inbred line，BIL）是一种新的群体分析法，即先连续回交再自交，可以获得性状互补的遗传资源（Ofner et al.，2016）。回交自交系与渐渗系相结合可对已有的QTLs进行精细定位，并找到具有强效应的候选基因（Fulop et al.，2016）。此外，由潘那利番茄构建的渐渗系已经分出了亚等位基因系（亚系），每个亚系携带的潘那利番茄基因片段都有定向的分子标记，而且亚系中所携带的潘那利番茄基因片段比渐渗系中的片段要小，更有利于候选基因的快速鉴定（Alseekh et al.，2013）。这些亚系可供研究机构使用，且已用于番茄果实化学成分效应基因的定位（Alseekh et al.，2015；Liu et al.，2016a，2016b）。目前设计构建的异源渐渗系库除潘那利番茄外，还包括醋栗番茄（Doganlar et al.，2003）、多毛番茄（Mon-forte et al.，2000；Finkers et al.，2007a，2007b）和类番茄茄（Canady et al.，2005）。

渐渗系也被用来解析杂种优势的遗传基础（Eshed et al.，1995）。杂种优势是指两个有差异的品种或种杂交产生的 F_1，在产量、生长发育以及育性等方面超过双亲的现象（Birchler et al.，2010）。杂种优势包含全基因组范围内的显性互补和特定位点的超显性遗传模型（Lippman et al.，2007）。在番茄渐渗系中已经鉴定出了一些性状的杂种优势 QTLs（Semel et al.，2006）。如加工番茄中一个特殊的 QTL，其在杂合状态时，番茄的收获指数、早熟性和代谢物含量（糖和氨基酸）均有提高（Gur et al.，2010，2011）；还有一个参与开花的自然突变 SFT 基因，该基因相当于一个超显性单基因，使得加工番茄杂交种的产量提高（Krieger et al.，2010）。

2.2.4 番茄抗病基因与 QTL

长期以来，番茄生产上普遍存在化学杀虫剂和杀菌剂过量使用的现象，化学药剂使用成本增加且防治效果有限，必须将遗传抗性和栽培管理技术结合起来才能实现农业可持续发展（Lefebvre et al.，2018）。然而，在长期驯化过程中，番茄栽培种的遗传多样性日益匮乏。多样性匮乏已成为培育抗性更强或耐受性更好的新品种的限制因素之一。研究人员曾在对番茄野生近缘种的筛选中发现了很多具有抗性性状的资源（Rick et al.，1995），其中在番茄野生种中鉴定到大约 40 个主效基因，这些基因对不同害虫和病原体引起的病害具有抗性，其中 20 个主效基因已经导入栽培番茄中（Ercolano et al.，2012）。研究表明，在秘鲁番茄（*S. peruvianum*）、多毛番茄、醋栗番茄和智利番茄（*S. chilense*）中拥有最丰富的抗性基因资源（Laterrot，2000）。对番茄种质资源抗病性进行系统筛选，将会发现新的抗性资源和新的抗性位点（主效抗性基因和抗性 QTLs）。

2.2.4.1 番茄抗病基因与 QTLs 的发掘

番茄中已经定位了 30 种主要病害的 100 多个抗性基因位点，与很多抗性基因或 QTLs 相关的分子标记也已有报道。截至目前，已经分离的主要抗性基因有 26 个，包括 *Asc-1*、*Bs-4*、*Cf-2*、*Cf-4*、*Cf-5*、*Cf-9*、*Hero*、*I*、*I-2*、*I-3*、*I-7*、*Mi-1.2*、*ol-2*、*Ph-3*、*pot-1*、*Prf*、*Pto*、*Sw-5*、*Tm-1*、*Tm-2*、*Tm-2^2*、*Ty-1*、*Ty-2*、*Ty-3*、*ty-5*、*Ve-1*（表 2-2）。番茄抗性位点的命名有明确规则：由 1～3 个斜体字母组成，第一个字母大写表明是显性基因，小写则是隐性基因，与数字之间用"-"隔开，数字表示目标病害抗性基因发现的顺序。有些情况下，在短横线后的一个数字之后用"."连接另一个数字，表示不同的等位基因，等位基因也可以用一个上标的数字或字母表示。

番茄抗性基因的显隐性。在目前已经克隆的番茄抗性基因中，大多数主效抗病基因都是显性基因，只有抗马铃薯 Y 病毒组［马铃薯 Y 病毒（PVY）和烟草蚀纹病毒（TEV）］的 pot-1、抗番茄黄化曲叶病毒（TYLCV）的 ty-5 和抗番茄白粉病的 ol-2 是隐性基因（Bai et al.，2008；Lapidot et al.，2015；Ruffel et al.，2005）。隐性基因 py-1 是调控由番茄棘壳孢引起的番茄木栓根腐病抗性的隐性抗病等位基因，虽有相关研究报道但目前还没有克隆（Doganlar et al.，1998）。

番茄不同病害抗性基因的研究进展不同。从大多数番茄病害中都已经发现了单个主效抗性基因，拥有多个主效抗性基因的只是一部分病害。如番茄斑萎病毒，已报道有 6 个显性抗性基因和 3 个隐性抗性基因（Foolad et al.，2012）；根结线虫也已鉴定出多个抗性基因。然而，广泛用于 MAS 育种的也只有 Sw-5 和 Mi-1.2，因为这两个基因较其他基因具有广谱抗性。仅有一小部分病害同时拥有主效抗性基因和抗性 QTLs，这些抗性基因和 QTLs 是通过所用的抗性供体和致病菌变种的鉴定分析以及环境条件鉴定得到的。

番茄抗性基因的等位性。番茄上已经克隆的抗性基因有些属于等位基因。如 6 号染色体上的 Ty-1 和 Ty-3（Verlaan et al.，2013），以及 9 号染色体上的 Tm-2 和 Tm-2^2（Lanfermeijer et al.，2005）。5 号染色体上的 Pto 和 Prf 紧密连锁，共同参与对番茄丁香假单胞菌（$Pseudomonas\ syringae$ pv. $tomato$）的识别（Salmeron et al.，1996a，1996b）。有些主效抗性基因属于抗性基因簇。如 1 号染色体的 Cf-4 和 Cf-9（Takken et al.，1999），以及 6 号染色体上的 Cf-2 和 Cf-5（Dixon et al.，1998）。

番茄抗性基因具有多效性。抗性基因通常对害虫、致病菌的种以及小种等具有专一性，但是，同样的基因对亲缘关系比较远的不同种害虫却鲜有抗性。Mi-1.2 也称 Meu，是一个特例，它不仅对由 3 个根结线虫属（$Meloidogyne$）［南方根结线虫（$M.\ incognita$）、花生根结线虫（$M.\ arenaria$）、爪哇根结线虫（$M.\ javanica$）］引起的根结线虫具有抗性，而且对蚜虫（$Macrosiphum\ euphorbiae$）、烟粉虱和木虱（$Bactericerca\ cockerelli$）也均有抗性（Casteel et al.，2007；Milligan et al.，1998；Nombela et al.，2003；Rossi et al.，1998；Vos et al.，1998）。

番茄上经分子鉴定的抗性基因见表 2-2。

表 2-2 番茄上经分子鉴定的抗性基因

位点（同名）	基因功能	发现的种（供体/来源）	携带该基因的资源材料	染色体	ITAG 基因名称	基因库编号	文献
Asc-1 (Asc)	LAG1	Solanum pennellii	LA716	T3	Solyc03g114600	AJ312131	Brandwagt et al., 2000
Bs-4	TIR-NB-LRR	Solanum lycopersicum	Money maker cultivar	T5	Solyc05g007850	AY438027	Schornack et al., 2004
Cf-2	LRR-RLP	Solanum pimpinellifolium	LA2244, LA3043	T6	Solyc06g008300	U42444	Dixon et al., 1996
Cf-4	LRR-RLP	Solanum habrochaites	LA2446, LA3045, LA3051, LA3267	T1	Solyc01g006550	AJ002235	Takken et al., 1998, 1999
Cf-5	LRR-RLP	Solanum lycopersicum	未知	T6	未知	AF053993	Dixon et al., 1998
Cf-9	LRR-RLP	Solanum pimpinellifolium	LA3047	T1	Solyc01g005160	AJ002236	Jones et al., 1994
Hero	CC-NB-LRR	Solanum pimpinellifolium	LA121	T4	Solyc04g008120	AJ457051	Ernst et al., 2002
I (I-1)	LRR-RLP	Solanum pimpinellifolium	PI79532	T11	Solyc11g011180	未知	Catanzariti et al., 2017
I-2	CC-NB-LRR	Solanum pimpinellifolium	PI126915	T11	Solyc11g071430	未知	Ori et al., 1997; Simons et al., 1998
I-3	SRLK-5	Solanum pennellii	LA716	T7	Solyc07g055640	KP082943	Catanzariti et al., 2015
I-7	LRR-RLP	Solanum pennellii	PI414773	T8	Solyc08g77740	KT185194	Gonzalez-Cendales et al., 2016
Mi-1.2 (Mi, Meu)	CC-NB-LRR	Solanum peruvianum	莫特尔品种，是大部分番茄砧木材料	T6	6 号染色体的几个同源区段	AF039682	Vos et al., 1998; Milligan et al., 1998; Nombela et al., 2001; Rossi et al., 1998; Casteel et al., 2007

(续)

位点(同名)	基因功能	发现的种(供体/来源)	携带该基因的资源材料	染色体	ITAG基因名称	基因库编号	文献
ol-2 (SlMlo1)	Loss-of-function mlo	Solanum lycopersicum	LA1230, KNU-12 cultivar	T4	Solyc04g049090	AY967408	Bai et al., 2008
Ph-3	CC-NB-LRR	Solanum pimpinellifolium	LA4285, LA4286, LA1269, (PI365957)L3708	T9	Solyc09g092280-Solyc09g092310	KJ563933	Zhang et al., 2013, 2014
pot-1	真核翻译启动因子 4E(eIF4E)	Solanum habrochaites	PI247087	T3	Solyc03g005870	AY723736	Ruffel et al., 2005; Piron et al., 2010
Prf	CC-NB-LRR	Solanum pimpinellifolium	LA2396, LA2458, LA3472	T5	Solyc05g013280	U65391	Salmeron et al., 1996
Pto	丝氨酸/苏氨酸蛋白激酶	Solanum pimpinellifolium	LA2396, LA2458, LA3472	T5	Solyc05g013300	U02271	Martin et al., 1993
Sw-5	CC-NB-LRR	Solanum peruvianum	PI128654	T9	Solyc09g098130	AY007367	Brommonschenkel et al., 2000
Tm-1	烟草花叶病毒 RNA 复制抑制子	Solanum habrochaites	PI126445	T2	Solyc02g062560	AB713135 AB713134	Ishibashi et al., 2007
Tm-2² (Tm-2ᵃ)	CC-NB-LRR	Solanum peruvianum	Craigella, GCR267	T9	Solyc09g018220	AF536201	Lanfermeijer et al., 2005
Tm-2	CC-NB-LRR	Solanum peruvianum	Craigella, GCR236	T9	Solyc09g018220	AF536200	Lanfermeijer et al., 2005

2　番茄的遗传和基因组资源

(续)

位点(同名)	基因功能	发现的种(供体/来源)	携带该基因的资源材料	染色体	ITAG基因名称	基因库编号	文献
Ty-1	DFDGD-Class RNA-Dependent RNA Polymerases RDR多聚酶	Solanum chilense	LA1969	T6	Solyc06g051170, Solyc06g051180, Solyc06g051190	—	Verlaan et al., 2013
Ty-2 (TYNBS1)	CC-NB-LRR	Solanum habrochaites	H9205, TY-Chie	T11	Solyc11g069660.1, Solyc11g069670.1	LC126696	Yamaguchi et al., 2018
Ty-3	DFDGD-Class RNA-Dependent RNA Polymerases RDR多聚酶	Solanum chilense	LA2279	T6	Solyc06g051170, Solyc06g051180, Solyc06g051190	AF272367	Verlaan et al., 2013
Ty-5	mRNA surveillance 检测因子	Solanum peruvianum	Tyking cultivar32 TY172	T4	Solyc04g009810	KC447287	Lapidot et al., 2015
Ve-1 (Ve)	RLP-type resistance protein 协助蛋白	Solanum lycopersicum	VFN8, Craigella GCR 151, PI303801	T9	Solyc09g005090	AF272367	Kawchuk et al., 2001; Fradin et al., 2009

注：抗性基因分类采用的是害虫和致病菌本身所属分类的拉丁名，每个基因都给出其相应的ITAG基因名称和基因库编号。

同时，番茄很多病害的主效抗性基因还没有找到，而有些已有主效抗性基因的抗性也由于致病菌的变异而丧失。鉴于此，研究者已经开始致力于数量性状抗性的开发，这种抗性作用只是阻碍病虫的发育而不是完全阻断它们。数量性状抗性部分抗性由 QTLs 调控。与主效基因抗性相比，数量型抗性在大多数情况下具有持久性和广谱性（Cowger et al.，2019）。此外，抗性 QTLs 在自然遗传资源中出现的频率要高于主效抗性基因。番茄基因组中已经定位了很多抗性 QTLs 特别是疫霉菌（Arafa et al.，2017；Brouwer et al.，2004；Brouwer et al.，2004；Foolad et al.，2008；Ohlson et al.，2018；Ohlson et al.，2016；Panthee et al.，2017；Smart et al.，2007）、番茄孢粉菌（Bai et al.，2003）、早疫病菌（*Alternaria solani*）（Foolad et al.，2002）、链格孢菌（*Alternaria alternata*）（Robert et al.，2001）、黄单胞菌属（*Xanthomonas* sp.）（Hutton et al.，2010；Sim et al.，2015）、青枯菌（*Ralstonia solanacearum*）（Carmeille et al.，2006；Mangin et al.，1999；Wang et al.，2013a，2013b）、灰霉病菌（Davis et al.，2009；Finkers et al.，2008；Finkers et al.，2007a，2007b）和黄瓜花叶病毒（*cucumber mosaic virus*，CMV）（Stamova et al.，2000）等的抗性位点。

番茄晚疫病的三个主效抗性基因中，*Ph-1* 的抗性已由于疫霉菌株系的进化而丧失，而携带 *Ph-2* 和 *Ph-3* 的植株也会受到疫霉菌进化株系的侵染，表现不完全抗性。由于番茄晚疫病这三个主效基因的抗性被攻克，人们已经着力在番茄近缘种和栽培种资源中寻找新的育种资源（Caromel et al.，2015；Foolad et al.，2014）。

当没有自然变异可以用于抗性育种，也没有外源 DNA 可以遗传转化时，可采用 TILLING 技术钝化植物中允许病原菌繁殖的显性敏感基因。利用这种方法已经在番茄中获得了对两种马铃薯 Y 病毒的抗性（Piron et al.，2010）。同样，利用 EcoTILLING 技术可以检测出同一基因的等位基因的自然变异，这种手段也已用于鉴定番茄中控制 TSWV 抗性的 *Sw-5* 的新变异（Belfanti et al.，2015）。

2.2.4.2 番茄抗病基因与 QTLs 的染色体分布

番茄中已定位的抗性位点尽管在基因组 12 条染色体上都有分布，但不是均匀分布的，其中有几个抗性基因分布的热点区域。同样地，主效抗性基因的分布也具有成双或成串存在的特点（Foolad，2007）。在 6 号和 9 号染色体上鉴定出了大量的抗性基因位点，这些位点或来自栽培种或来自野生近缘种。6 号染色体上的主效抗性基因：*Mi-1.2* 抗根结线虫，*Ol-1*、*Ol-3*、*Ol-4*、*Ol-5* 和 *Ol-6* 抗番茄白粉病（新番茄粉孢菌，*Oidium neolycopersici*），*Cf-2* 和 *Cf-5* 抗番茄叶霉病（番茄叶霉菌，*Cladosporium fulvum*），*Ty-1* 和 *Ty-3* 抗番茄黄化曲叶病毒，*Am* 抗苜蓿花叶病毒（*alfalfa mosaic virus*）。6 号染色体上的抗性 QTLs 主要抗青枯病和抗番茄花叶病毒（*tomato mottle virus*，ToMoV）（Agrama et al.，2006）。同样，9 号染色体上也富含抗性基因簇。如 *Tm-2* 和 *Tm-2²* 抗番茄花叶病毒（Pillen et al.，1996）；*Frl* 位于近着丝点区，抗番茄茎腐根腐病（Vakalounakis et al.，1997）；*Sw-5* 抗番茄斑萎病毒（Stevens et al.，1995）；*Ph-3* 位于端粒附近，抗番茄晚疫病（Chunwongse et al.，2002）；*Ve* 位于另一端粒附近，抗黄萎病菌（*Verticillium dahliae*）（Kawchuk et al.，2001）。

2.2.4.3 番茄抗病基因与 QTLs 的分子基础

番茄中多数抗性性状是由单显性基因控制的，抗性基因编码的蛋白直接或间接识别害虫和致病菌蛋白并使其失活，进而启动植物的防御反应。也有部分抗性性状由单隐性基因控制，通常用小写字母表示，如 *pot-1* 和 *ol-2*。隐性抗性等位基因主要可以使致病菌发生功能性丧失或敏感性缺失，进而使病菌在植物体内的发育和繁殖受阻碍。相反，相应的感病等位基因则促进和加速致病菌在宿主体内的发育。很多主效抗性基因已经通过正向遗传学和图位克隆法得以分离克隆，而且大多数显性基因都编码保守的 NB-LRR 蛋白。这种抗性基因的保守分子结构域（NBS-LRR、R-genes、RLP、RLK 等）已被用于在同种或近缘种中寻找和分离同源基因，以发现和分离新的等位基因。如 *Sw-5* 和 *Mi* 具有相同的结构域，*Cf* 系列基因也一样。近年来，采用抗性基因序列捕获 RenSeq［resistance（R）gene sequence capture］技术，根据从茄科已经鉴定出的 260 个具有 NBS-LRR 结构域的基因设计诱饵，已经从番茄 Heinz1706 基因组中成功鉴定出 105 个新的具有富含亮氨酸的重复结合位点（NBS-LRR）的核苷酸序列，在醋栗番茄 LA1589 基因组中发现了 355 个新的 NBS-LRR。这些研究完善了这些种中具有编码 NBS-LRR 结构的 R 基因（Andolfo et al.，2014）。番茄主效抗性基因遗传图谱见图 2-1。

除了这些主效抗性基因外，在番茄中还鉴定出了很多在病害防御反应中被激活的基因。其中有些基因对植物和致病菌之间的反应具有专一性，而有些基因可以参与多个植物和致病菌之间的互作。如脂肪酶样蛋白 EDS1 能够参与由 Cf-4 蛋白和 Ve 蛋白诱导的防御过程，同样地，Prf 蛋白、I-2 蛋白和 Bs-3 蛋白与 RAR1 蛋白、SGT1 蛋白和 HSP90 蛋白之间都有互作。此外，通过转录组分析，发现了几个参与茉莉酸或水杨酸信号通路调控的基因，其中有些基因的作用与抗性 QTLs 相当。

截至目前，番茄中还没有克隆出对抗病性具有决定作用的QTLs，目前获得的植物数量性状抗性位点相当于具有部分抗性作用分子的集合。有些基因可能参与识别病原相关分子模式（pathogen-associated molecular pattern，PAMP），负责识别和基础防御；有些基因参与防御信号传导；有些基因参与调控植物防御素合成。而R基因的弱效应等位基因、调控发育表型的基因或其他基因还没有鉴定出来（Poland et al.，2009）。

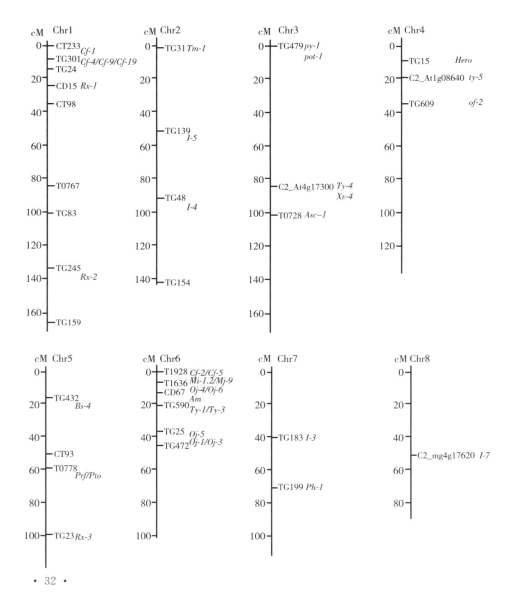

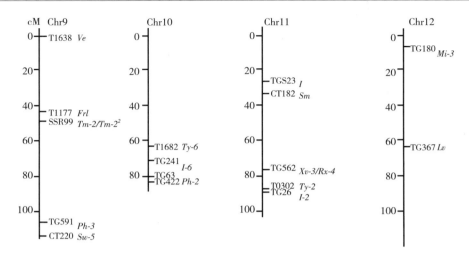

图2-1 番茄主效抗性基因遗传图谱

注：标记名称和遗传距离根据番茄 SGN-EXPEN2000 图谱整理，基因位置来源于 Foolad（2007）、Foolad 等人（2014）、Lee 等人（2015）、Bai 等人（2018）、Gill 等人（2019）和 Sharma 等人（2019）的文章。当文章和 EXPEN2000 图谱中没有共同标记时，通过番茄染色体 SL2.50 鉴定最近的标记，再进行连锁标记核苷酸序列比对来确定相对位置。遗传距离（cM）在染色体左侧显示。

2.3 番茄的基因组资源

2.3.1 番茄参考基因组序列

番茄基因组信息极大地促进了人们对现代番茄进化历史和基因结构的理解。番茄基因组测序项目作为国际茄科基因组测序计划的组成部分，于2003年3月在美国华盛顿启动，由来自中国、法国、西班牙、意大利、美国、英国、荷兰、日本、韩国和印度10个国家的科学家研究团队实施（Mueller et al.，2005）。由于番茄与3 000多个种存在高度宏共线性关系和微共线性关系，因此，番茄成为茄科作物基因组测序中的首选实验材料。起初，项目采用的是传统的双脱氧测序法。双脱氧测序法为一代测序技术。为了保障质量、降低成本，以饱和遗传标记为基础的细菌人工染色体（bacterial artificial chromosome，BAC）BAC-by-BAC 测序法得以建立，即在番茄基因组中的基因富集区筛选种子 BACs 用于测序。整个过程比较缓慢，限制了测序的进程。随着二代测序技术的出现，这些问题迎刃而解，进度得到显著提升。

2012年，第一个番茄基因组序列发表。测序所用材料为栽培番茄自交系

Heinz 1706 及其近缘野生种醋栗番茄 LA1589（The Tomato Genome Consortium，2012）。在番茄基因组中，重组基因和转录本主要位于常染色质区而非异染色质区。与之不同，叶绿体插入基因和保守的 MicroRNA 基因在整个基因组几乎是均匀分布的（The Tomato Genome Consortium，2012）。番茄基因组与茄科其他作物，如辣椒、茄子、马铃薯和烟草具有高度的同源性，与拟南芥和高粱相比，番茄中很少发现在古老时期发生的具有高拷贝、全长的长末端重复转座子。番茄基因组注释显示，番茄基因组中总共有 34 727 个编码蛋白的基因，其中 30 855 个基因得到 RNA 测序数据支持。番茄的基因、转录子、重复序列和 sRNA 在染色体上的排布与马铃薯非常相似。编码蛋白的基因中，有 8 615 个基因在番茄、马铃薯、拟南芥、水稻和葡萄中是共有的。番茄中预测保守的 sRNAs 共有 96 个，可分为 34 个家族，其中有 10% 在植物中高度保守。马铃薯基因组约有 8% 与番茄不同，其中有 9 个大的倒位和几个小的倒位（The Tomato Genome Consortium，2012）。茄科经历了一次古老的和多次近代连续的基因组三倍体化过程，使得人们对于番茄遗传多样性的狭窄及其引起的遗传瓶颈有了基本的认识（The Tomato Genome Consortium，2012）。

番茄基因组序列第一版发表后，又经过进一步的完善、修正，并利用新的序列数据和 RNAseq 数据进行重新注释。目前基因组版本更新为 SL3.0，注释版本为 ITAG3.2。

2.3.2 番茄基因组重测序

二代测序技术的出现使得大范围的基因组测序成为可能（Goodwin et al.，2016）。番茄基因组参考序列发布不久，2014 年，具有抗逆性的野生潘那利番茄的基因组序列发布（Bolger et al.，2014）。鉴定结果显示，潘那利番茄具有极端抗旱性和不同寻常的形态学结构，其中很多抗逆相关候选基因位点已经完成定位。结果发现，潘那利番茄 LA716 和栽培番茄 M82 之间存在大量的基因表达差异，这种差异由启动子和（或）编码区序列多态性所致。通过长读测序平台与 Illumina 测序相结合，2017 年，潘那利番茄及其他相关种基因组也完成了重测序和数据注释（Usadel et al.，2017）。随后，大量番茄材料以及野生近缘种完成了测序。结果表明，具有自交不亲和性的野生种杂合性最高，而具有自交亲和性的种杂合性较低。同时，对 360 份番茄材料的基因组重测序结果分析揭示了番茄育种的演化历史（Lin et al.，2014），发现番茄驯化和改良的过程主要由两组相互独立的 QTLs 在发挥作用，这些 QTLs 导致果实增大。

驯化过程中对果实增大起作用的主效 QTLs 有 5 个，分别是 $fw1.1$、$fw5.2$、$fw7.2$、$fw12.1$ 和 $lcn12.1$；然后，有 17 个主效 QTL 实现了番茄果实的第二次改良和增大，包括 $fw1.1$、$fw2.1$、$fw2.2$、$fw2.3$、$lcn2.1$、$lcn2.2$、$fw3.2$、$fw3.2$、$fw5.2$、$fw7.2$、$fw9.1$、$fw10.1$、$fw11.1$、$fw12.1$、$fw11.3$、$fw12.1$ 和 $lcn12.1$。同时，研究还检测出了主效基因 $SlMYB12$ 的几个独立突变体，能将番茄的果色由红色变为粉色，粉色番茄在亚洲地区很受欢迎。研究还发现，连锁累赘与野生片段渗入相关联（Lin et al.，2014）。

此后，低度重测序或基因分型测序已经普遍应用于番茄材料分析和研究。截至目前，约有 900 个番茄材料完成了重测序，测序深度从低到高不等（The Tomato Genome Consortium，2012；Causse et al.，2013；Bolger et al.，2014；Lin et al.，2014；The 100 Tomato Genome Sequencing Consortium，2014；Tieman et al.，2017；Ye et al.，2017；Tranchida-Lombardo et al.，2018）。这些基因组信息资源均可从茄科基因组网站获得，利用这些基因组信息将加速智能型番茄新品种的育种进程。

近期，学界对 725 份在进化关系和地理分布方面具有广泛代表性的番茄材料进行了泛基因组分析，通过与参考基因组序列比对发现了 4 873 个新基因（Gao et al.，2019），其中 272 个可能是污染基因，并已从 $Heinz\ 1706$ 参考基因组中移除。分析发现，番茄在驯化和改良过程中存在大量的基因丢失事件和对基因和启动子的高强度负向选择事件。在番茄驯化过程中，共检测到 120 个有益基因和 1 213 个不利基因；在番茄改良过程中，共检测到 12 个有益基因和 665 个不利基因。

抗病基因特别容易丢失或被负向选择。通过基因富集分析发现，在番茄驯化和改良过程中，防御反应对不利基因的富集程度最高；而在改良过程中，并没有发现有益基因显著富集的家族。一个 $TomLoxC$ 基因的启动子区的稀有突变在驯化选择过程中被发现。同时含有稀有和普通 $TomLoxC$ 等位基因的番茄材料中，这两个基因在果实橙色期的表达量高于仅含有一个基因的纯合体（现代番茄材料中通常仅含有一个基因）。综合以上研究结果，泛基因组学研究将为今后生物学发现和育种提供理论基础（Gao et al.，2019）。

2.4 番茄的 SNP 标记

2.4.1 番茄 SNP 的发现

在主要作物中，单核苷酸多态性（single nucleotide polymorphism，SNP）

是最丰富的分子标记，它能分布在基因组的任何区域，包括基因编码序列、非编码序列以及基因之间的间隔区域。然而，只有基因编码区的非同义 SNP 才能改变蛋白质的氨基酸序列。非编码区的 SNP 也能影响基因的表达，但其作用机制不同（Farashi et al.，2019）。SNP 数量巨大，可直接通过基因分型测序（genotyping-by-sequencing，GBS）或材料的重测序来获得（Catchen et al.，2011）。以二代测序为基础的各种新技术已经在主要作物中普遍应用，加快了农艺性状相关基因的鉴定与分离（Le Nguyen et al.，2018）。基因分型测序有多种方法，包括至少 13 种简化基因组测序（reduced-representation sequencing，RRS）和 4 种全基因组重测序（whole-genome resequencing，WGR）途径（Scheben et al.，2017）。其中，基于转录组的 RNA 测序和外显子组测序就是一种重要的简化基因组测序途径（Haseneyer et al.，2011；Scheben et al.，2017）。在事先没有基因组序列信息的情况下，利用转录组测序结果就可以进行表达分析（Wang et al.，2010）。

自从番茄参考基因组序列发布以来，学界通过对不同番茄材料进行全基因组重测序，获得了数以百万计的 SNP，实现了对番茄基因组的全覆盖（Bolger et al.，2014；Lin et al.，2014；Menda et al.，2014；The 100 Tomato Genome Sequencing Consortium，2014；Tieman et al.，2017；Ye et al.，2017；Zhu et al.，2018）。野生番茄 SNP 数量已超过 1 000 万个，是栽培番茄材料的 20 倍以上（The 100 Tomato Genome Sequencing Consortium，2014）。有了参考基因组序列，就可以只针对目标染色体区域进行 SNP 筛选。如 Ranc 等人（2012）利用 90 个番茄材料组成的核心资源在不同定位密度下对 2 号染色体进行重测序，共获得了覆盖 2 号染色体的 DNA 片段 81 个，发现了 352 个 SNP。

2.4.2 番茄 SNP 微阵列芯片

SNP 微阵列芯片是另一种常用的经济有效的基因分型途径，如茄科协同农业计划（solanaceae coordinated agricultural project，SolCAP）（Hamilton et al.，2012；Sim et al.，2012b）、生态系统基因组中心（centre for biosystems genomics，CBSG）联盟（Víquez-Zamora et al.，2013）及多样性序列技术（diversity arrays technology，DArT）（Pailles et al.，2017）。然而，基于 RNAseq 的 SNP 多态性芯片，如 SolCAP 和双酶切限制性位点相关 DNA 测序 ddRAD-Seq（restriction site-associated DNA sequencing）（Arafa et al.，2017）具有很大的局限性，基因表达具有时空性，会因组织和时间的不同而变化，因此在 RNA 片段化建立文库的过程中就会产生大量偏差（Wang et al.，2009），且这种 SNP 对编码区的覆盖率较低（Scheben et al.，2017）。SNP 微阵列芯片已经广泛用于对不同番茄材料进行基因分型（Sim et al.，2012a；Viquez-Zamora et al.，2013；Ruggieri et al.，2014；Sauvage et al.，2014；Blanca et al.，2015；Bauchet et al.，2017a，2017b；Pailles et al.，2017；Albert et al.，2016b）。

2.4.3　番茄基因型填充

当拥有大量且多样的参考序列数据时，SNP 的密度可通过基因型填充而显著增加（Guan et al.，2008；Halperin et al.，2009；Iwata et al.，2010；Marchini et al.，2010；Pasaniuc et al.，2012；Browning et al.，2016；Das et al.，2016；Wang et al.，2018）。在模式植物和人类中，已经有一些参考序列芯片适用于基因型填充，如 1000 基因组项目（The 1000 Genomes Project Consortium）、人类 UK10K 项目（The UK10K Consortium）（Danecek et al.，2015）、3000 水稻基因组项目（3000 Rice Genome Project）（McCouch et al.，2016）、拟南芥 1001 基因组联盟（the 1001 Genomes Consortium in Arabidopsis thaliana）。与全基因组测序相比，番茄中 SNP 芯片的标记密度非常低，存在大量基因组缺口（Sauvage et al.，2014；Bauchet et al.，2017b；Zhao et al.，2019）。通过基因型填充，SNP 的数量可以增加 30 倍，基因组缺口将得到连接和填充，基因组的覆盖率也会大大提高（图 2-2）（Zhao et al.，2019）。

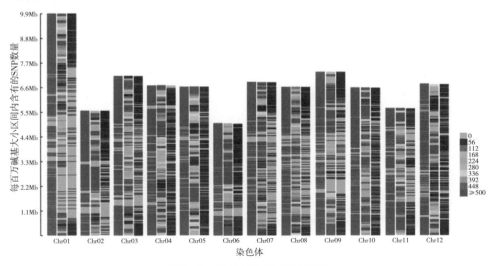

图 2-2　番茄材料的 SNP 密度

注：左、中、右列表示基因型填充前后参考序列集的 SNP 密度（Zhao et al.，2019）

2.5 番茄的遗传多样性分析

分子标记在现代育种中发挥着非常重要的作用（Ramstein et al.，2018），为人们对番茄遗传多样性的认识提供了新的视角（Bauchet et al.，2012）。总体来讲，现代栽培番茄各种分子标记的多样性低于野生种，如限制性片段长度多态性（RFLP）（Miller et al.，1990）、扩增片段长度多态性（AFLP）（Suliman-Pollatschek et al.，2002；Park et al.，2004；Van Berloo et al.，2008；Zuriaga et al.，2009）、随机扩增多态 DNA（RAPD）（Grandillo et al.，1996a；Archak et al.，2002；Tam et al.，2005；Carelli et al.，2006；El-hady et al.，2010；Meng et al.，2010；Length，2011）、简单重复序列标记（SSR）（Suliman-Pollatschek et al.，2002；Jatoi et al.，2008；Mazzucato et al.，2008；Albrecht et al.，2010；Meng et al.，2010；Sim et al.，2010；Zhou et al.，2015）、简单重复区间序列标记（ISSR）（Vargas-Ponce et al.，2011；Shahlaei et al.，2014）和单核苷酸多态性（SNP）（Blanca et al.，2012；Sim et al.，2012a；Lin et al.，2014；The 100 Tomato Genome Sequencing Consortium，2014）。

全基因组测序技术检测出了数以百万计的 SNP，其中野生种的 SNP 超过 1 000 万种，是大多数驯化栽培种的 20 倍以上（The 100 Tomato Genome Sequencing Consortium，2014），这为寻找番茄驯化和改良过程中丢失的遗传多样性提供了线索。番茄野生种和栽培种全基因组测序比较表明，在驯化过程中，大约 1% 的番茄基因组都经历了高强度的纯化选择（Sahu et al.，2017）。在表达水平上，驯化引起现代番茄与野生醋栗番茄之间差异表达的基因有 1 729 种，其中包含 17 个基因簇，驯化也使得一些基因的调控路径明显富集，如碳水化合物代谢和表观遗传调控（Sauvage et al.，2017）。

樱桃番茄遗传多样性介于野生种和栽培种之间（Ranc et al.，2012；Xu et al.，2013；Zhang et al.，2017），其连锁不平衡也处于两者之间（Sauvage et al.，2014；Bauchet et al.，2017a）。因此，樱桃番茄可作为有益的中间桥梁，将野生种中一些基因和强遗传负荷用来弥补栽培种在遗传多样性低和形态多样性高方面与野生种的差距。分子标记也能够将遗传和形态多样性联系在一起以探究番茄的起源。Blanca 等人（2012）利用 *SOLCAP* 基因分型 SNP 阵列对 272 份具有遗传和形态多样性的番茄材料进行表型分型发现，樱桃番茄在遗传和形态多样性两方面均居于野生醋栗番茄与现代栽培番茄之间。该研究还发现，樱桃番茄和野生醋栗番茄分布于完全不同的生态和气候区域，其群体结构和基于气候带划分的地理位置之间存在明确的相关性（Blance et al.，2012）。

2.6 番茄中已克隆的基因和 QTL

番茄是育种中应用单突变体数量最多的作物之一（Grandillo et al.，2016；Rothan et al.，2019）。在 SNP 发现之前，由于栽培番茄遗传多样性有限，用于连锁作图的群体通常是由栽培种与其近缘野生种之间通过杂交构建的分离群体（Foolad，2007；Foolad et al.，2012）。自从有了分子标记，这些分离群体就成了构建高密度遗传连锁图谱的高效工具（Tanksley et al.，1992）并用于发掘 QTLs。截至目前，采用限制性片段长度多态性（RFLP）、简单重复序列标记（SSR）和单核苷酸多态性（SNP）等多种分子标记在不同连锁群体中已经挖掘出数百个 QTLs，涉及农艺性状、形态性状和品质相关性状等（Grandillo et al.，1996b；Tanksley et al.，1996；Fulton et al.，1997；Bernacchi et al.，1998a，1998b；Chen et al.，1999；Grandillo et al.，1999；Fulton et al.，2000；Monforte et al.，2000；Saliba-Colombani et al.，2001；Causse et al.，2002；Doganlar et al.，2003；Van Der Knaap et al.，2003；Fridman et al.，2004；Baldet et al.，2007；Foolad，2007；Jiménez-Gómez et al.，2007；Cagas et al.，2008；DalCin et al.，2009；Sim et al.，2010；Ashrafi et al.，2012；Haggard et al.，2013；Kinkade et al.，2013）。

然而，在已经发掘出的 QTLs 中，仅有一小部分被克隆并进行了功能验证（Bauchet et al.，2012；Rothan et al.，2019）。番茄上第一个通过图位克隆的基因是 *Pto*，借助 RFLP 标记完成了定位和克隆，该基因对丁香假单胞菌株系具有抗性（Martin et al.，1994）。接着，利用同一个 RFLP 图谱，同一个基因家族的另一个成员 *Fen*（Martin et al.，1994）很快被克隆。此后，学界利用 RFLP 标记鉴定和克隆了不同的抗性基因，如 *Cf-2*，富含亮氨酸重复蛋白，抗番茄叶霉病菌（Dixon et al.，1996）；*Prf* 是另外一个对丁香假单胞菌番茄株系具有抗性的基因（Salmeron et al.，1996）；*Ve* 抗黄萎病，编码表面受体（Kawchuk et al.，2001）等抗性基因。除了 RFLP 外，其他分子标记也相继被开发并用于抗性基因的鉴定。如来源于醋栗番茄抗晚疫病疫霉菌的 *Ph-3*，就是借助酶切扩增多态性序列（cleaved amplified polymorphic sequences，CAPS）或插入缺失 InDel（insert/deletion）标记克隆获得的（Zhang et al.，2014）；利用 CAPS 和序列特异性扩增区域（sequence characterized amplified region，SCAR）标记完成了番茄黄化曲叶病毒病抗性基因 Ty-2 的定位（Yang et al.，2014）。

利用分子标记还鉴定和克隆了一些参与发育过程的重要基因和 QTLs。其中，fw2.2 调控番茄果重的主效 QTLs 就属首批。研究者利用 CAPS 标记定位并克隆了位于 2 号染色体上的单个候选基因 ORFX（Frary et al.，2000），该基因主要通过表达调控影响果实大小，对编码蛋白的结构和序列没有影响（Nesbitt et al.，2002）。近年来，功能得到鉴定的主效 QTLs 包括 fw3.2（细胞色素 P450 基因）（Chakrabarti et al.，2013）和 fw11.2（细胞大小调控因子）（Mu et al.，2017）。据报道，OVATE 是果重紧密相关的主效 QTLs，是一个负向调控基因，能使番茄果实形状呈梨形（Liu et al.，2002），其他调控果实形状的基因还有反转录转座子调节基因 SUN（Xiao et al.，2008）、果实心室数量基因 fas（Huang et al.，2011）和 lc（Muños et al.，2011）。

番茄果实中富含多种对健康有益的营养物质，如糖、有机酸、氨基酸和挥发性物质。然而，培育高营养和风味浓郁的番茄品种仍然是育种中一个主要的挑战（Tieman et al.，2012；Klee et al.，2013；Klee et al.，2018；Zhao et al.，2019）。Lin5 是一个调控番茄果实中糖含量的主效 QTL，20 年前已经完成了克隆（Fridman et al.，2000）。在不同遗传背景和环境条件下，与栽培种相比较，野生种中的等位基因能够增加果实中葡萄糖和果糖的含量（Fridman et al.，2000），同时，这个基因在番茄中的表达模式与在马铃薯和拟南芥中的表达模式相似（Fridman et al.，2003）。最近有研究表明，一种 SWEET 蛋白定位于质膜上的葡萄糖转运子，对葡萄糖和果糖的比例具有重要作用（Shammai et al.，2018）。糖和有机酸含量的平衡对消费者喜好至关重要（Tieman et al.，2017）。最近，有人克隆了一个调控苹果酸含量的主效 QTL（Sl-ALMT9），对应铝活化苹果酸转运蛋白 9（Ye et al.，2017）。新的研究发现，这个 QTL 也可能调控番茄果实中柠檬酸盐的含量（Zhao et al.，2019）。尽管关于糖和有机酸调控 QTLs 的功能分析较少，但是已有的结果对研究人员了解糖和有机酸的调控机制非常重要。此外，已有人鉴定出了几个参与调控挥发物含量的基因（Tieman et al.，2006；Tikunov et al.，2013；Klee，2010；Klee et al.，2018）。

2.7　番茄新的基因和 QTL 鉴定

Lin 等人（2014）通过对番茄 F_2 群体极端混池全基因组重测序，鉴定出了很多与果实重量相关的 QTLs，包括 $fw2.1$、$fw2.2$、$fw2.3$、$lcn2.1$、$lcn2.2$、$fw9.1$、$fw9.3$、$fw11.1$、$fw11.2$ 和 $fw11.3$。利用 F_2 群体的表型极端混池进行全基因组重测序，将所得结果与已有参考基因组序列进行比对，进而鉴定出与目标表型相关的 SNP（Garcia et al.，2016）。这种分析方法对突变体特别是 EMS 突变体的鉴定非常有效（Garcia et al.，2016）。

然而，连锁群体的遗传多样性受杂交双亲的影响。为了克服这种局限，可采用多亲本高阶世代互交（multi-parent advanced generation intercross，MAGIC）群体。MAGIC 已经在拟南芥（Kover et al.，2009）、水稻（Bandillo et al.，2013）、小麦（Huang et al.，2012；Mackay et al.，2014）、蚕豆（Sallam et al.，2015）、高粱（Ongom et al.，2017）和番茄（Pascual et al.，2015）等作物上应用。番茄上第一个 MAGIC 群体是用 8 个重测序的番茄自交系杂交构建而成，在这个群体中没有明显的群体结构。用 MAGIC 群体构建的连锁图谱要比双亲本群体的大 87% 以上，由此鉴定出了一些影响果实品质的主效 QTLs（Pascual et al.，2015）。近年来，这种 MAGIC 群体法已用于水分胁迫和盐胁迫下 QTLs 的鉴定，并已通过鉴定获得了很多胁迫特异性 QTLs（Diouf et al.，2018）。

2.8 番茄全基因组关联分析

2.8.1 番茄全基因组关联分析的条件

关联作图就是利用由不相关材料构成的群体，研究某一已知表型基因与遗传标记之间的相关性。如果遗传标记覆盖整个基因组，就称之为全基因组关联分析（GWAS）。全基因组关联分析技术最先在人类研究中发展形成，显示出了其在人类疾病分析方面的作用（Klein et al.，2005），并很快在主要作物上得以应用（Brachi et al.，2011；Luo，2015；Liu et al.，2019）。番茄上首次应用全基因组关联分析技术是用来鉴定与 $fw2.2$ 果重主效 QTL 相关的 SNP。然而，此次研究中，在由 39 个樱桃番茄组成的小群体中，并没有找到正相关的 SNP（Nesbitt et al.，2002）。

为了在番茄中有效使用全基因组关联分析技术，需要采用不同的分子标记对不同类型番茄的连锁不平衡现象（linkage disequilibrium，LD）进行评估。总体来看，LD 在栽培番茄中大约为 20Mbs，大于野生番茄；樱桃番茄介于两者之间（Van Berloo et al.，2008；Mazzucato et al.，2008；Sim et al.，2010；Ranc et al.，2012；Xu et al.，2013；Sauvage et al.，2014；Zhang et al.，2016a，2016b；Bauchet et al.，2017a）。这些结果表明，番茄驯化和育种过程中产生了遗传多样性丢失。将樱桃番茄与栽培番茄、野生番茄混合渗入有助于减轻 LD，克服现代番茄品种关联作图效率低的问题（Ranc et al.，2012）。然而若 LD 平均程度高，覆盖全基因组所需分子标记的最低数量就少。例如，Xu 等人（2013）仅用了 121 个 SNP 和 22 个 SSR 对 188 个番茄材料进行关联作图，成功鉴定出了 132 个与 6 个果实品质性状显著相关的位点。在大量 SNP 没有开发之前，主要用 SSR 进行全基因组关联分析。Zhang 等人（2016a，2016b）用 182 个 SSR 标记对 174 个不同基因型群体（包括 123 个樱桃番茄和 51 个传统老品种）进行果实品质性状全基因组关联分析，共鉴定出 111 个与 10 个性状显著关联的标记。之前鉴定出的许多主效 QTLs 也都位于关联标记的位置及其附近。在此基础上，作者进一步拓展了表型范围，同样鉴定出了很多与挥发性物质（Zhang et al.，2016a，2016b）、糖和有机酸（Zhao et al.，2016）显著相关的标记，其中一些与用 SNP 进行同样性状全基因组关联分析的结果一致（Sauvage et al.，2014；Bauchet et al.，2017b；Tieman et al.，2017；Zhao et al.，2019）。

利用番茄基因组参考序列（The Tomato Genome Con-sortium，2012）可以获得数以百万计的 SNP，进而可用于目标性状遗传多态性鉴定。例如，调控番茄果实粉色的基因 *SlMYB12*，就是基于重测序对 231 个番茄材料进行全基因组关联分析鉴定获得的（Lin et al.，2014）。Lin 等人（2014）还进一步鉴定出了该基因的 3 个隐性等位基因和几个 SlMYB12 蛋白结构突变体，这些研究成果对粉果番茄育种帮助很大。

全基因组测序技术目前仍然十分昂贵，尤其是对大规模群体而言，极大地限制了该项技术的广泛应用。SNP 微阵列芯片技术克服了这一局限性（Hamilton et al.，2012；Sim et al.，2012b）。Sauvage 等人（2014）利用 SolCAP 微阵列分型技术对 163 个由大果番茄、樱桃番茄和野生番茄构成的群体进行基因分型，获得了 5 995 个高质量的 SNP，然后采用多位点混合模型（multi-locus mixed model，MLMM）（Segura et al.，2012）对 36 种在 2 个发育时期高度相关的代谢物质进行全基因组关联分析，鉴定出了 44 个与果实不同代谢物质相关的候选基因位点（Sauvage et al.，2014），其中一个功能未知的候选基因位于 6 号染色体上，该基因与苹果酸含量密切相关。这种相关关系在后来不同群体的全基因组关联分析的整合分析（meta-analysis）中得到了进一步验证（Bauchet et al.，2017b；Tieman et al.，2017；Ye et al.，2017；Zhao et al.，2019），并确认该基因编码铝激活苹果酸转运蛋白 Sl-ALMT9

(Al-activated malate transporter 9)（Ye et al.，2017）。根据 3 个群体的全基因组关联分析进一步发现，该基因与番茄果实中柠檬酸的含量也具有显著相关关系，表明其在调控番茄有机酸方面具有重要作用（Zhao et al.，2019）。事实上，铝激活苹果酸转运蛋白是一个植物特异蛋白家族，对植物根部组织及其功能发挥具有重要作用（Delhaize et al.，2007）。

Bauchet 等人（2017b）采用 SolCAP 和 CBSG 微阵列分型技术对 300 个番茄材料进行基因分型，获得了 11 012 个高质量 SNP 并将其用于全基因组关联分析，同时采用多位点混合模型和多性状混合模型 2 种分析模型（Korte et al.，2012），共鉴定出 79 个与番茄果实代谢物显著关联的标记位点，包括 13 个初生代谢产物和 19 个次生代谢产物，其中有 2 个位点与果实酸度和苯丙酸类物质含量显著相关（Bauchet et al.，2017b）。此外，他们还利用同一群体进行了农艺性状及其 QTLs 鉴定，如 $fw2.2$ 和 $fw3.2$，通过试验分析发现，在栽培种和野生种之间存在着同时具有两者不同单倍型的中间类型（Bauchet et al.，2017a）。类似品质性状的全基因组关联分析在其他群体中也有报道（Ruggieri et al.，2014；Zhang et al.，2016a，2016b）。

随着全基因组测序技术的迅速发展和成本的不断降低，大批量番茄材料的基因测序成为可能。例如，Tieman 等人（2017）对 231 个新的番茄材料进行重测序，将其结果与之前已经测序的 245 个材料相结合，构建了包含 476 个基因组序列的数据库，用于对风味相关代谢物的全基因组关联分析，包括 27 种挥发物、总固形物、葡萄糖、果糖、柠檬酸和苹果酸，检测到与 20 个性状显著关联的位点 251 个，其中有 2 个位点与葡萄糖和果糖显著关联，对应的 2 个主效 QTLs 为 $Lin5$ 和 $SSC11.1$。结合选择分析表明，糖含量与果实重量之间呈负相关关系，这种关系的形成，可能主要是因为在大果番茄的驯化和改良过程中高糖等位基因丢失（Tieman et al.，2017）。此外，还鉴定出一些有益于番茄挥发物含量积累的候选基因，如 $Solyc09g089580$ 调控愈创木酚和水杨酸甲酯含量。通过对 3 个香叶基丙酮和 6-甲基-5-庚烯-2-酮显著关联位点分析发现，调控风味物质的等位基因组合在驯化和育种过程中逐渐丢失（Tieman et al.，2017）。

2.8.2　番茄全基因组关联的 Meta 分析

在对番茄相同性状开展的不同全基因组关联分析结果中，仅有部分关联性显著的位点能够被鉴定出来，这就表明不同研究之间存在强异质性，即不同全基因组关联分析的遗传效应存在非随机方差。导致异质性的主要因素有群体结构、连锁不平衡、表型测量方法、环境因素、基因型分型方法、基因型与环境互作等（Evangelou et al.，2013）。全基因组关联分析的 Meta 分析作为一种新的分析手段，可以正确处理不同全基因组关联分析之间的异质性，将不同全基因组关联分析结果结合起来进行综合分析。

Zhao 等人（2019）报道了从 3 个不同番茄群体获得全基因组关联分析的 Meta 分析结果（Sauvage et al.，2014；Bauchet et al.，2017b；Tieman et al.，2017）。分析中共用了 775 个番茄材料和 2316117 个 SNP，通过基因型填补，鉴定出了 305 个与糖含量、有机酸、氨基酸和风味相关挥发物显著关联的标记。通过对 5 个与果糖和葡萄糖相关联的位点分析发现，随着野生等位基因数量的增加，糖含量显著增加，同时表明驯化和改良过程影响了柠檬酸和苹果酸含量，特别是苹果酸主效 QTLs，*Sl-ALMT9* 与柠檬酸显著相关，1 号染色体上鉴定出的另一个苹果酸转运子也与柠檬酸含量显著相关。该研究同时还鉴定出了很多与风味相关挥发物显著关联的标记位点，对 6 个显著关联位点的进一步研究表明，与樱桃番茄相比，现代番茄风味欠佳，这主要是因为在现代番茄果实中，人们喜好的风味物质含量低，而人们不喜好的风味物质含量高（Zhao et al.，2019）。

2.9 番茄非生物胁迫耐受性的遗传解析

2.9.1 番茄基因型与环境互作（G×E）的遗传调控

非生物胁迫对番茄的影响在前面章节中已有介绍。对于非生物胁迫的响应，野生番茄与栽培番茄之间存在很大差异，不同栽培番茄材料之间也存在较大差异。探索番茄对环境胁迫响应差异遗传机制的试验研究表明，番茄属丰富的遗传多样性是番茄对非生物胁迫响应差异的遗传决定因素。

番茄中可利用的遗传资源极其丰富，野生番茄和栽培番茄中都蕴藏着丰富的遗传资源。因此，在两个种群中筛选遗传多样性丰富的种类，可以减少栽培番茄种群中由于长期、高强度的农艺性状定向筛选导致的多样性严重丢失（Lin et al.，2014）。尽管如此，在气候条件和生长条件多样化的环境下，栽培番茄种群中也保存了大量的环境响应基因，这些基因决定了番茄的环境适应性程度。通过对番茄不同种内试验群体观察发现，番茄对非生物胁迫的适应性，除了基因的作用，还有大量的基因型与环境互作的影响（Villalta et al.，2007；Mazzucato et al.，2008；Albert et al.，2016a；Diouf et al.，2018）。

野生番茄中之所以蕴含着大量与非生物胁迫耐受性相关的特异基因，是由于野生番茄长期以来对于自身生长环境和典型有害生境的适应。例如，多毛番茄和潘那利番茄对冷害（Bloom et al.，2004）、干旱和盐胁迫（Bolger et al.，2014）的耐受性显著强于栽培番茄。野生番茄中的耐受性基因和栽培番茄中胁迫响应基因的遗传多样性为番茄环境智能型育种奠定了资源基础。番茄环境智能型育种前途光明。

随着高密度遗传图谱的出现，针对番茄对非生物胁迫的遗传特性的研究不断深入。Grandillo 等人（2013）和 Grandillo 等人（2016）对番茄中响应不同非生物胁迫的 QTL 进行了汇总报道，表 2-3 中仅列举了过去 10 年内鉴定出的 QTL。这些 QTL 的获得采用了不同类型的群体和不同的作图方法，所用方法涵盖了目前已有的植物遗传作图方法。基于胁迫响应机制的复杂性，研究筛选出了一些表型性状，用于评价番茄对非生物胁迫的响应以及耐受性。例如，Kazmi 等人（2012a，2012b）在水分胁迫、冷胁迫、盐胁迫和高温胁迫下，采用番茄种子质量性状鉴定出了与种子发芽力相关的 QTLs。通过这种方法，在胁迫条件下鉴定出了约 90 个与种子质量相关的 QTLs。Arms 等人（2016）和 Asins 等人（2017）在水分胁迫和氮胁迫条件下，分别利用 sub-NIL（Arms et al.，2016）和 130 个 F_{10} RIL（Asins et al.，2017）作图群体完成了对几个生理指标的定位。Rosental 等人（2016）研究盐胁迫条件下番茄种子的代谢物变化，在 72 个由潘那利番茄 LA716 和栽培种 M82 构建的近等基因系中鉴定出几个相关 QTLs。最近有人通过对水分胁迫和对照条件下基因表达数据分析，鉴定出了几个与水分胁迫互作的 eQTL（表达数量性状位点）（Albert et al.，2018）。通过顺式和反式调控 eQTL 的差别，揭示番茄在水分胁迫下的表达调控模式以及由其引起的基因型与环境的互作机制。QTL 分析与表达数据相结合有助于鉴定出胁迫响应候选基因，筛选出适宜的遗传标记，用于胁迫适应性分子标记，辅助选择（MAS）。

然而，大多数研究都是用农艺性状代替生理指标或代谢性状去评价非生物胁迫的影响。应针对不同的育种目标，选择和定义不同的胁迫指标，这样一来，利用这些胁迫指标获得的 QTLs 就可直接应用于育种。

截至目前，大多数关于非生物胁迫响应的研究都是在单一胁迫下进行的，获得的遗传位点也只是参与对给定非生物胁迫的响应。未来的研究应该在多个逆境因子组合胁迫的条件下，在全基因组范围内研究对组合胁迫的响应位点，目前植物中这样的例子几乎没有（Davila Olivas et al.，2017）。

过去 10 年内发表的番茄非生物逆境胁迫 QTLs 见表 2-3。

表 2-3 过去 10 年内发表的番茄非生物逆境胁迫 QTLs

处理	个体数量	标记	胁迫处理	处理时期	杂交设计	表型	QTL数量	参考文献
低温胁迫（CS）								
CS	83 RILs	865 SNP	12 ℃	种子萌发期	双亲种间杂交	种子质量	12	Kazmi et al.，2012
CS	146 RILs	120 SSR	11 ℃	种子萌发期	双亲种间杂交	发芽率	5	Liu et al.，2016a，2016b
CS	146 RILs	120 SSR	2 ℃，48 h	4~5 叶期	双亲种间杂交	冷害	9	Liu et al.，2016a，2016b
高温胁迫（HT）								
HT	192 F2	106 AFLP	最低>25 ℃，最高<40 ℃	定植至试验结束	双亲种间杂交	坐果情况	6	Grilli et al.，2007
HT	83 RILs	865 SNP	35~36 ℃	种子萌发期	双亲种间杂交	种子质量	16	Kazmi et al.，2012
HT	180 F2	96 SNP	白天温度 31 ℃，夜间温度 25 ℃	从第一花序出现开始	双亲种间杂交	繁殖性状	13	Xu et al.，2017a，2017b
HT	98 F8 RILs	727 SNP	37 ℃	种子萌发期	双亲种间杂交	耐热性、热抑制、热休眠	9	Geshnizjani et al.，2018
HT	160 F2	62 RAPD ISSR AFLP	白天温度 37.2 ℃，夜间温度 24.7 ℃	整个生长期	双亲种间杂交	产量、果实品质、繁殖性状	21	Lin et al.，2010
盐胁迫（SS）								
SS	123 RILs	156 SSR SCAR	125 mmol/L NaCl	定植后 15 d 至试验结束	双亲种间杂交	砧木诱导的生理指标、营养生长情况	57	Asins et al.，2010
SS	52 ILs	—	150 mmol/L NaCl	7 叶期开始处理 21 d	双亲种间杂交	植株结构、过氧化物含量	71	Frary et al.，2010
SS	52 ILs	—	150 mmol/L NaCl	处理 15 d	双亲种间杂交	植株结构、营养生长情况	225	Frary et al.，2011
SS	78 ILs	—	700 mmol/L NaCl，70 mmol/L $CaCl_2$	定植后 4 d 内	双亲种间杂交	存活表型	4	Li et al.，2011
SS	90 ILs	—	700 mmol/L NaCl，70 mmol/L $CaCl_2$	定植后 4 d 内	双亲种间杂交	存活表型	6	Li et al.，2011

2 番茄的遗传和基因组资源

（续）

处理	个体数量	标记	胁迫处理	处理时期	杂交设计	表型	QTL 数量	参考文献
SS	100 RILs	134 SSR, SCAR	75 mmol/L NaCl	定植后15 d至试验结束	双亲种间杂交	砧木诱导的生理指标、营养生长情况	2	Asins et al., 2010
SS	83 RILs	865 SNP	-0.3 MPa NaCl, -0.5 MPa NaCl	种子萌发期	双亲种间杂交	种子质量	32	Kazmi et al., 2012
SS	124 RILs	2059 SNPs	8.94 dS/m	定植后10 d内	双亲种间杂交	产量、果实品质、生物量	54	Asins et al., 2015
SS	72 ILs	—	6 dS/m	定植后至试验结束	双亲种间杂交	种子重量、种子发芽率、代谢物	131	Rosental et al., 2016
SS	253 MAGIC RILs	1 345 SNP	3.7 dS/m, 6.5 dS/m	定植后至试验结束	MAGIC（种内杂交）	果实品质、植株结构和营养生长、物候期、产能	35	Diouf et al., 2018

水分胁迫（WD）

处理	个体数量	标记	胁迫处理	处理时期	杂交设计	表型	QTL 数量	参考文献
WD	75 ILs	—	1 000 m² 灌溉水量 30 m³	定植后至试验结束	IL（种间杂交）	果实品质、植株结构和营养生长、物候期、产能	114	Gur et al., 2011
WD	83 RILs	865 SNP	-0.3 MPa PEG, -0.5 MPa PEG	种子萌发期	双亲种间杂交	种子质量	23 (19)	Kazmi et al., 2012
WD	119 RILs	679 SNP	40% 蒸散量	定植后至试验结束	双亲种间杂交	果实品质、植株结构和营养生长、物候期、产能	36	Albert et al., 2016a
WD	141 small-fruit accessions	6 100 SNPs	40% 蒸散量	定植后至试验结束	GWAS-panel	果实品质、植株结构和营养生长、物候期、产能	100	Albert et al., 2016b
WD	18 sub-NILs	10 (SNP; SCAR; CAP)	33% 蒸散量	定植后至试验结束	NIL（种间杂交）	生理性状、植株结构	2	Arms et al., 2016

(续)

处理	个体数量	标记	胁迫处理	处理时期	杂交设计	表型	QTL数量	参考文献
WD	117 F7 RILs	501 SNP	49%蒸散量	定植后至试验结束	双亲种间杂交	模型参数	8	Constantinescu et al.,2016
WD	241 MAGIC RILs	1 345 SNP	50%蒸散量	定植后至试验结束	MAGIC(种内杂交)	果实品质、植株结构和营养生长、物候期、产能	22	Diouf et al.,2018
WD	124 RILs	501 SNP	60%蒸散量	定植后至试验结束	双亲种间杂交	果实品质、植株结构和营养生长、物候期、产能	23	Albert et al.,2018
WD	124 RILs	501 SNP	60%蒸散量	定植后至试验结束	双亲种间杂交	274 个基因表达量	103 个 eQTLs	Albert et al.,2018
其他胁迫								
过氧化胁迫	83 RILs	865 SNP	300 mm H_2O_2	种子萌发期	双亲种间杂交	种子质量	17	Kazmi et al.,2012
缺氮胁迫	130 F10 lines	1899 SNP	0.1 mmol/L NH_4^+,1 mmol/L NO_3^-	定植后至第一序坐果	双亲种间杂交	营养生长、叶片氮含量、木质部汁液中激素含量	40	Asins et al.,2017

注:每个研究所用的基因型数量、杂交群体构建以及标记的数量和类型均在表中。"胁迫处理"和"处理时期"栏表明胁迫的水平和胁迫处理的周期;"表型"栏列举了 QTL 和关联分析所用的表型性状,表型性状通常分为种子质量(发芽力)、果实品质(可溶性固形物含量、维生素 C、pH、硬度、有机酸)、植株结构和营养生长(直径、叶长、高度、干物质含量、比叶面积、生物量)、物候期(开花时间、成熟时间)、生产效率(产量、果实重量、果实数量)、生理性状(WUE)、模型参数(细胞壁最大延展性、膜电导率、糖活性吸收、膜反射、花梗导度、可溶性糖浓度、果实干重、果实含水量、木质部导度)。"—"表示具体信息未知。

当作物受到非生物胁迫时,基因型与环境互作通常就会发生。育种家们应对基因型与环境互作的策略通常有两种:①培育适应某种特殊目标环境的优良品种;②培育对环境条件具有广适性的品种。第一种策略可以在可预知的环境条件下(或可控制的环境条件下)获得高产;第二种策略的产量水平虽会在不可预知的环境条件下低于最优水平,但在实际应用中这种策略会更有效。为此,植物遗传学家对基因型与环境互作相关表型可塑性的遗传调控问题进行探究,并已经在主要作物中鉴定出了不同可塑性 QTLs。例如 Kusmec 等人(2017)认为,玉米中大多数控制对不同环境可塑性的基因与控制一般性状变异的基因明显不同,并提出可将产量性状和稳定性性状进行同步选择。在番茄研究中，有人利用种间杂交群体，在水分胁迫和盐胁迫条件下，鉴定获得了番茄可塑性 QTLs（Albert et al.，2016a；Diouf et al.，2018）。进一步拓宽对不同胁迫条件的环境范围才有可能获得对多个胁迫响应的基因，才有助于环境智能型番茄品种的培育。

2.9.2 番茄嫁接对胁迫耐受性的影响

嫁接作为一种能克服土传病害、增强作物对各种非生物胁迫抵抗能力的有效途径（King et al.，2010）已被广泛应用于植物，特别是果树和蔬菜作物生产中。将优良品种嫁接到抗性砧木上是应对极端土壤环境胁迫的有效选择之一，这既可以避免野生基因渗入引起连锁累赘的副作用，又可以获得对非生物胁迫的多基因抗性。嫁接中需要特别注意接穗与砧木的亲和性，因为两者是否亲和对嫁接效果至关重要。番茄中不同的嫁接试验研究表明，接穗与砧木的相互作用表现在以下几个方面：嫁接可引起果实品质组成、植株生长势、植株激素水平以及最终产量的变化（Kyriacou et al.，2017）。这些研究结果一方面强调测试接穗和砧木亲和性的必要性，另一方面要求搞清楚嫁接对育种目标性状的影响，这样才能在胁迫环境下有效利用砧木达到预期效果。

番茄中已经培育出了不同的砧木群体，包括来自樱桃番茄与两个野生近缘种醋栗番茄和克梅留斯基番茄种间杂交获得的后代群体（Estañ et al.，2009）。对这些群体在盐胁迫（Albacete et al.，2009；Asins et al.，2010，2015，2013）和氮缺乏（Asins et al.，2017）胁迫下的表型研究表明，在盐胁迫下，嫁接可以诱导叶部激素含量和离子浓度的改变，这些激素含量和离子浓度与植株营养生长和产量相关。盐胁迫下砧木的介导作用表现为多基因效应特点，受到 7 号染色体上不同 QTLs 位点调控，这些位点与 2 个 HTK 候选基因相关，参与离子转运和细胞稳定态调控。同时，有报道指出，在盐胁迫条件下，嫁接可以提高番茄产量或是保持番茄产量不变，也存在生理性病害脐腐病发病率提高、果实成熟延迟等不足。

砧木诱导的激素状态改变能提高番茄水分利用率（Cantero-Navarro et al.，2016）。Nawaz 等人（2016）综合论述了园艺作物中嫁接对离子积累的效应，同时强调嫁接作为一种应对作物土壤极端环境的有效技术途径，还需要在基因和表型两个方面，对砧木×接穗×环境之间相互作用开展进一步研究。

将感病品种嫁接到抗性砧木上可以有效应对土壤中的生物和非生物胁迫。也有观点认为，嫁接可以通过增强 RNA 沉默效应提高作物对病毒的抗性（Spano et al.，2015）。目前面临最大的挑战，就是培育出能够同时抵抗生物胁迫和非生物胁迫的砧木。

2.10 番茄组学分析

2.10.1 番茄代谢组学分析

代谢组学在番茄多样性研究方面具有重要作用（Schauer et al.，2005；Fernie et al.，2011）。代谢组学分析可以对已知代谢途径的代谢物开展进一步分析研究（Tieman et al.，2006），也可以对未知代谢途径的新代谢物进行鉴定分析（Tikunov et al.，2005），以促进人们对番茄果实组分的认识，进而加速番茄的品质育种进程（Fernie et al.，2009；Allwood et al.，2011）。代谢组学分析已经用于对果实组成成分的全面分析。目前已经鉴定出相关 QTLs（mQTL）的代谢物，包括非挥发性物质（糖、色素）和挥发性物质等（Bovy et al.，2007；Klee，2010，2013；Klee et al.，2018）。这些研究主要利用种间杂交群体，特别是潘那利番茄（Alseek et al.，2015，2017）和克梅留斯基番茄（Do et al.，2010；Ballester et al.，2016）渐渗系和种内杂交组合（Saliba-Colombani et al.，2001；Causse et al.，2002；Zanor et al.，2009）。此外，代谢组学分析也被用于研究番茄植株与蓟马的相互作用（Mirnezhad et al.，2010）。

2.10.2 番茄转录组学分析与 eQTL 定位

有研究通过果实发育过程转录组变化分析，发现了不同发育阶段基因表达上的关键性变化（Pattison et al.，2015；Giovanonni et al.，2017；Shinozaki et al.，2018），并通过分离群体对这些变化的遗传调控进行了分析研究（Ranjan et al.，2016；Coneva et al.，2017）。另外，鉴定分析不同环境条件下基因表达的多样性也是认识基因型与环境互作的重要步骤。Albert 等人（2018）鉴定出了一些能响应水分胁迫的 eQTLs，发现在正常灌溉和水分胁迫两种不同条件下，叶片和果实中转录组之间存在很大差异。他们同时研究了等位基因在 F_1 中的特异性表达（ASE），表明等位基因偏离 1∶1 分离比例，大部分的基因变异受到 ASE 与水分供应量互作的显著影响，其中约 80%响应水分胁迫的基因都是通过大量交互作用来调控的。

2.10.3 番茄多组学分析

Prudent 等人（2011）指出，代谢组学和转录组学相结合可以找到果实组分遗传控制的线索。Zhu 等人（2018）将基因组、转录组和代谢组三方面的数据进行整合，开展了多组学研究，鉴定出 3 526 个与 514 种代谢物显著关联的位点，其中 351 个位点与未知代谢物相关联。通过基因组学与转录组学相关性分析，共鉴定出 2 566 个顺式数量性状区间（cis-eQTL）和 93 587 个反式数量性状区间（trans-eQTL）；经代谢组学和转录组学之间严格的多重校正测验，共鉴定出 232 934 个表达量与代谢物含量关联的位点，涉及 820 种化学物质和 9 150 个基因。经过 3 种组学的整合分析，共鉴定出 13 361 个代谢物 SNP 基因三方相关关系（metabolite-SNP-gene），包括 371 种代谢物、970 个 SNP 和 535 个基因。通过选择分析发现了 168 个驯化区间和 151 个改良区间，分别占番茄基因组的 7.85% 和 8.19%，共有 4 095 个和 4 547 个基因分别定位在驯化区间和改良区间内。此外，共鉴定出 46 个与甾类物质糖苷生物碱相关联的位点，有 5 个显著关联位点定位在驯化或改良区间内。研究结果同时表明，抗性基因的渗入会引起一些代谢物的显著性差异。

2.10.4 番茄 MicroRNA 与表观遗传修饰

表观基因组学是指特定时间、特定细胞中每个基因组位点表观遗传标记的完整集合（Taudt et al., 2016）。表观遗传学是与遗传学相对应的概念。表观遗传学是研究在基因的核苷酸序列不发生改变的情况下基因表达的可遗传变化，它是遗传学的分支学科。表观遗传学研究基于非基因序列改变所导致的基因表达水平变化，如 DNA 甲基化和染色质构象变化等。表观基因组学是在基因组水平上对表观遗传学改变的研究。表观遗传标记分为 6 类，包括 DNA 修饰、组蛋白修饰、染色质变异、核小体定位、RNA 修饰和非编码 RNAs、染色质构象变化及其相互作用（Stricker et al., 2017）。利用当今技术可以在全基因组范围内对表观基因组学变异进行高分辨率测量，目前已在人类、老鼠、玉米、番茄、拟南芥和大豆上取得了阶段性成果（Taudt et al., 2016；Giovannoni et al., 2017）。

番茄表观基因组学研究主要集中在果实成熟和发育的分子调控方面（Gallusci et al., 2016；Giovannoni et al., 2017），其中组蛋白转录后修饰具有重要作用，包括赖氨酸残基的磷酸化作用、甲基化作用、乙酰化作用和单泛素化作用（Berr et al., 2011）。在拟南芥中，组蛋白转录后修饰作用在植物发育和胁迫适应性等很多方面都有体现（Ahmad et al., 2010；Mirouze et al., 2011）。研究人员已经在番茄中鉴定出至少 9 种 DNA 甲基化转移酶和 4 种 DNA 去甲基化酶（Gallusci et al., 2016），也明确了几种组蛋白修饰因子在新鲜果实中的表达模型，如组蛋白去乙酰化酶、组蛋白乙酰转移酶和组蛋白甲基转移酶（Gallusci et al.，2016）。番茄多梳蛋白抑制复合体 2（PRC2）中的 *SlEZ1* 和 *SlEZ2* 作用不同：抑制 *SlEZ1* 的表达可以引起花和果实的形态变化（HowKit et al.，2010）；抑制 *SlEZ2* 的表达只影响果实的形态变化，如质地、颜色和耐储性（Boureau et al.，2016）。这些结果表明表观遗传调控在很多生物过程中具有重要作用。

番茄表型与表观遗传突变相关联的例子不多。无色不成熟位点（*Cnr*）是番茄果实成熟调控网络中的一个主要元件（Eriksson et al.，2004），Manning 等人（2006）在 *Cnr* 的 SBP-box 启动子区鉴定出了一个自然发生的甲基化表观遗传突变，Quadrana 等人（2014）鉴定出了一个调控果实维生素 E 含量的表观遗传突变。为了明确表观基因重塑是否参与了果实成熟过程，Zhong 等人（2013）研究发现，番茄果实在甲基化转移酶抑制因子 5-氮杂胞嘧啶核苷作用下提早成熟。在番茄基因组中已鉴定出的甲基化区域有 52 095 个，占番茄基因组的 1%。值得注意的是，已经鉴定出的去甲基化区域均位于大量成熟基因的启动子区。此外，表观基因组学不是静止不变的，而是随果实成熟在不断变化（Zhong et al.，2013）。Shinozaki 等人（2018）绘制了一张番茄果实发育和成熟过程的高分辨率转录组时空图谱，鉴定出了一些组织特异性的成熟相关基因，如 *SlDML2*。结合其他分析得知，时空甲基化作用在番茄果实发育和成熟过程中发挥着重要作用（Shinozaki et al.，2018）。

Lü 等人（2018）以 7 种呼吸跃变型果实（苹果、香蕉、甜瓜、木瓜、梨、桃和番茄）和 4 种非呼吸跃变型鲜果（黄瓜、葡萄、草莓和西瓜）为材料，通过对果实 ENCODE 数据库中的 361 个转录组、71 个染色质区域、147 个组蛋白和 45 个 DNA 甲基化数据的功能性元件分析，在呼吸跃变型果实中鉴定出了 3 种转录反馈调控路径，调控乙烯依赖型果实的成熟。在非呼吸跃变型果实中，*H3K27me3* 协同开花调控因子 *FLOWERING LOCUS C* 和花同源异形基因 *AGAMOUS* 的沉默，通过限制成熟基因及其同源序列调控干果和非乙烯依赖型果实的成熟，并且这种作用具有保守性。

MicroRNA（miRNA）是表观遗传调控的一种类型，是一类由 20～24 个核苷酸构成的内源非编码小 RNA。这些 MicroRNA 能够裂解转录物和抑制转录后的翻译（Chen，2005，2009；Rogers et al.，2013；Sanei et al.，2015），在转录和转录后水平调控中具有非常重要的作用。MicroRNAs 的编码基因包含 TATA-box 结构域和转录因子结合域，受特异性转录因子的调控（Xie et al.，2005；Megraw et al.，2006；Rogers et al.，2013；Yu et al.，2017）。

MicroRNA 在人类（Calin et al.，2006；Mendell et al.，2012；Cui et al.，2017b；Hill et al.，2018）、动物（Ambros，2004；Rajewsky，2006；Grimson et al.，2008）和植物（Rogers et al.，2013；Won et al.，2014；Sanei et al.，2015；Cui et al.，2017a；You et al.，2017；Yu et al.，2017）的很多生物过程中具有重要作用，包括生理、发育、防卫和（适应）环境变化等。如今，人们揭示了一些切割复合体核心元件的调控机制，如 *DICER-LIKE1*（*DCL1*）和 *HYPONASTIC LEAVES1*（*HYL1*）（Manavella et al.，2012；Cho et al.，2014；Zhang et al.，2017）；鉴定出了促进 MicroRNA 合成加工和减少 MicroRNA 含量水平的蛋白，如帽结合蛋白 80（CAP-BINDING PROTEIN 80，CBP80）、帽结合蛋白 20（CAP-BINDING PROTEIN 20，CBP 20）和稳定蛋白 1（STABILIZED1，STA1）等（Gonatopoulos-Pournatzis et al.，2015；Yu et al.，2017）；发现 CDC5、NOT2、Elongator 和 DDL 蛋白（Yu et al.，2008；Wang et al.，2013a，2013b；Zhang et al.，2013；Fang et al.，2015）能够减少成熟和未成熟 MicroRNA 的积累。截至目前虽已发现了很多参与 MicroRNA 合成、降解和活性调控的生物过程，但关于这些过程的亚细胞定位还知之甚少（Yu et al.，2017）。

根据番茄基因组测序结果预测有 96 个保守的 MicroRNA 基因，其中已鉴定出的 MicroRNA 有 34 种，10 种在番茄和马铃薯中高度保守（The Tomato Genome Consortium，2012）。番茄果实发育过程中 MicroRNA 的鉴定分析已成为研究热点（Moxon et al.，2008；Zuo et al.，2012；Gao et al.，2015），sRNA 中的研究热点主要是 21-24nt 的 sRNA（Mohorianu et al.，2011；Zuo et al.，2012；Gao et al.，2015）。很多成熟相关基因的转录因子都受相应 MicroRNA 家族的调控，如 *miR156/157*、*miR159*、*miR160/167*、*miR164*、*miR171* 和 *miR172* 家族（Moxon et al.，2008；Karlova et al.，2013；Zuo et al.，2013）。MicroRNA 前体基因也受很多反式因子的调控（Rogers et al.，2013）。乙烯有可能参与 MicroRNA 及其前体基因的调控，如 *TAS3-mRNA*、*miR156*、*miR159*、*miR160*、*miR164*、*miR171*、*miR172*、*miR390*、*miR396*、*miR4376* 和 *miR5301*（Gao et al.，2015）。成熟抑制子 RIN 通过 MicroRNA 和乙烯对相关基因转录后调控，实现对番茄果实成熟相关基因的调控。此外，乙烯还可以通过调节 mRNA 的丰度来调控 MicroRNA（Gao et al.，2015）。研究发现，MicroRNA 可以特异性地诱导番茄对生物和非生物胁迫的响应，因此，可以利用 MicroRNA 来改变番茄的适应性（Liu et al.，2017）。在新鲜果实发育和成熟

过程中，表观基因组学调控具有重要作用，然而在果实成熟以及环境胁迫条件下，表观基因组学的动态研究仍需进一步加强（Giovannoni et al.，2017）。

2.11 番茄数据库

访问数据库是获得各种有关番茄研究数据的重要途径。番茄研究者们的愿望之一就是将遗传和基因组学数据汇集起来形成一个自由开放的数据库，目前这个愿望已经实现。大家熟知的茄科基因组网站（Solanaceae Genome Network）包含了茄科几个物种的数据资源。该数据库资源丰富，其中有第一张 RFLP 遗传图谱，也有有关标记、基因和 QTL 的信息资源，还有目前的基因组浏览器，利用该浏览器可以找到已经发现的基因组和 SNP 资源。除此之外，其他几个数据库对于番茄遗传学家也很有帮助，其中包含了遗传资源、突变体或是基因表达等信息资源（表 2-4）。

表 2-4 番茄遗传学和基因组学主要数据库

名称	网址	内容范围
茄科基因组协作组（Solanaceae Genome Network，SGN）	https://solgenomics.net	茄科基因组学中心（基因组序列、位点、表型等）
番茄遗传资源中心（Tomato Genetic Resource Center，TGRC）	https://tgrc.ucdavis.edu	位于 UC Davis 的 Charles Rick 番茄遗传资源中心
番茄突变体库（Tomato Mutants Archive，Tomatoma）	https://tomatoma.nbrp.jp	Microtom 突变体及其基因组数据档案
番茄表达数据库	https://ted.bti.cornell.edu	基因表达分析结果
番茄表达图册（Tomato Expression Atlas）	https://tea.solgenomics.net	高分辨率基因表达图谱
番茄转录组数据库（Tomexpress）	https://tomexpress.toulouse.inra.fr login	RNAseq 数据库
番茄 EFP 浏览器（Tomato EFP browser）	https://bar.utoronto.ca/efp_tomato/cgi-bin/efpWeb.cgi	番茄基因表达综述
番茄基因组和代谢组数据库（Solcyc）	https://solgenomics.net/tools/solcyc/index.pl	基因组学和代谢途径数据库

3 番茄智能型基因组设计育种

3.1 番茄传统育种

番茄是一种自花授粉作物，第一代的番茄品种都是地方品种（系）。番茄集约化育种始于20世纪30年代的美国。作为一种自花授粉作物，在相当长的时间里，包括系谱选择法和回交选育法的重组育种法是番茄育种的主要途径。早期育种通过这些传统方法先后将野生种中的优良基因转入栽培种中，使栽培种的抗病性显著提升，同时也改进了果实的硬度和其他品质性状（Bai et al.，2007）。采用轮回选择法能有效提升番茄果实糖含量和果实大小，同时能打破这两个性状之间的负相关关系（Causse et al.，2007a，2007b）。

番茄杂种优势育种始于20世纪70年代，自此F_1杂交种逐渐取代了纯系品种，但番茄的杂种优势在产量方面不是很显著。在杂交种育种中发现，番茄果实形状与果实大小具有同质性，而且几个显性抗性基因可以同时组合在一个杂交种中，一个杂交种可以同时聚合6～8个抗病基因。F_1种子生产是番茄杂交育种的一个重要环节，为了方便F_1种子生产，雄性不育性利用研究受到重视。一套核隐性雄性不育基因虽已被发现，但目前尚未被用于商业生产。功能性雄性不育基因，由位置不育突变体基因 *ps2* 调控，主要表现为花药不能自然开裂，因此无法散粉，具有开发应用前景（Atanassova，1999）。尽管如此，携带不育基因的材料通常伴随着选育难度大的问题，面对频繁的番茄品种更新换代，F_1种子生产普遍在劳动力成本较低的国家通过人工授粉来完成。

3.2 番茄标记辅助选择育种

目前，标记辅助选择（marker-assisted selection，MAS）已在番茄育种中普遍应用。由于番茄很多重要位点已经作图定位并开发有相应的分子标记，在利用MAS时，育种家关注的是与目标性状表达相关的基因组区域。MAS选择的效率和复杂性因目标性状的遗传特性（单基因还是多基因）而不同。就单基因性状而言，标记辅助回交法（marker-assisted backcross，MABC）就是最直接有效的途径，而对于多基因控制的性状则需要采用不同的策略。

3.2.1 番茄标记辅助回交法

标记辅助回交法适用于单个基因的选择，其原理很简单。先鉴定具有与目标基因紧密连锁的分子标记，利用这个标记对供体导入基因进行有效检测，这个过程称为前景选择。同时，利用受体亲本其他位点的分子标记对相应位点进行检测选择，以加速受体亲本其他位点的基因型回归，这个过程称为背景选择。背景选择不仅要用携带导入基因的染色体上的标记，同时还要用其他染色体上的标记。基于对供体染色体进行背景选择的标记，可以鉴定出在该基因一侧或两侧发生重组的个体，以控制伴随导入基因的供体基因组片段的长度（Young et al.，1989）。经过三代 MABC 选择，基因纯合度显著高于传统选择方法。通过比较，要达到同样的基因纯度，传统方法至少要多进行 2 个世代以上（Hospital et al.，1992）。目前，番茄许多重要基因已经被定位（有了图谱）或克隆，学者们也开发出了有益等位基因的特异标记（Rothan et al.，2019）。番茄育种者普遍使用分子标记进行多个单基因性状，如抗病性或果实特异性状等的导入聚合。随着基因分型技术费用的降低，加快大规模植株筛选选择进程成为可能。

3.2.2 番茄 QTL 标记辅助选择

数量性状通常由几个 QTL 调控，而每个 QTL 的个体效应有限且彼此不同，故必须同时操控影响目标性状的几个 QTL 才有效。基于这类性状遗传的复杂性，需要根据目标性状 QTL 的数量、效应范围和特性以及有利等位基因的来源等采用不同的 MAS 策略。

就单基因性状而言，如果调控性状的 QTL 数量少且都来自同一个亲本，若要同时导入同一个优系中，那么 MABC 就是最有效的方案。Hospital 等人（1997）通过研究确定了在前景选择中检测调控性状 QTL 所需分子标记的最适数量和位置，给出了在几百个单株构成的群体中同时可以检测出 QTL 的最大数量。平均来讲，每个 QTL 至少需要 3 个标记，经过几代筛选，才能降低该 QTL 位点非供体基因型存在的风险。然而，每个世代保障目标基因型出现的最小个体数量需根据如下参数确定：①置信区间长度；②标记数量；③QTL 数量。同时转入 4~5 个及以上 QTLs 的可能性比较小，除非拥有非常大的群体，或 QTL 定位准确性非常高。

通过对番茄果实品质性状的 QTL 定位分析（Saliba-Colombani et al., 2001；Causse et al., 2001），已获得了几个品质相关的 QTL 簇。由于大多数用于品质改良的有益基因来自樱桃番茄（亲本系），为了将樱桃番茄中 5 个对果实品质影响最大的基因组片段导入 3 个轮回亲本中（Lecomte et al., 2004b），Lecomte 等人专门建立了一套 MABC 育种方案。群体大小规模要保障将 5 个片段成功导入每个轮回亲本中，而且 MAS 育种方案可以在不进行选择的条件下，将来自供体的非目的染色体片段大小控制在期望水平以下。筛选分别携带 1～5 个 QTLs 的植株，并以其为对象研究每个 QTL 的独立效应和几个 QTLs 的组合效应，结果表明大多数 QTLs 在携带一个供体片段的受体材料中都可以检测出来，除此之外，也可以检测出一些新的 QTLs（Causse et al., 2007a, 2007b）。与亲本材料相比，获得导入片段的受体材料的果实品质能得到改良，并具有进一步提升的潜力。相反，具有目标基因型材料的果实重量总是低于预期值，这是一些非预期 QTLs 作用的结果。由于这些对果实重量具有负效应的等位基因 QTLs 一开始并没有被鉴定出来，其效应在近等基因系群体选择过程中往往被掩盖或忽略了。

3.2.3 番茄高阶回交选择

高阶回交（advanced backcross）QTL 分析是一种育种策略，可以同时发现和转导新的等位基因，是一种专门用于将供体材料中有价值的 QTLs 等位基因同时导入优良自交系并进行鉴定的育种方法（Tanksley et al., 1996）。QTL 分析一般要在回交高代进行（BC_3 或 BC_4），通常前期进行的负向选择主要是为了降低供体有害等位基因存在的概率。而利用 BC_3/BC_4 群体选择可以减小供体片段的大小，减少连锁累赘，同时限制上位效应，节约时间。与建立携带同样 QTL 的 NIL 所需时间相比，高代回交 QTL 分析法所需的时间更短（Fulton et al., 1997）。Tanksley 等人采用这种策略在醋栗番茄（Tanksley et al., 1996）、多毛番茄（Bernacchi et al., 1998a）、秘鲁番茄（Fulton et al., 1997）、潘那利番茄（Eshed et al., 1996）和小花番茄（*S. parviflorum*）（Fulton et al., 2000）5 个野生种中筛选出了正效应等位基因。他们在不可直接利用的种质资源中鉴定出了大量具有应用潜力的有益等位基因，这些基因可以同时导入优良的栽培番茄品系中，一些重要的糖基转移基因则可以用于加工番茄的品质改良，由此挖掘出异源种质资源中所隐藏的利用价值（Bernacchi et al., 1998b, Tanksley et al., 1996）。

3.2.4 番茄聚合设计育种

Hospital 等人(1997)建议当导入 QTLs 的数量较多时采用聚合设计育种。为了降低高阶回交选择的强度,先采用 MABC 对 QTLs 进行逐一检测,然后将筛选出的单株个体进行相互杂交,以达到将有益等位基因 QTLs 聚合到同一个基因型材料中的目的。当有益等位基因来源不同时,Van Berloo 等人(1998)建议采用指数法筛选出能用于杂交的重组自交系。选择指数最优的植株进行杂交,以便将尽可能多的有益 QTLs 聚合到单个基因型材料中,利用这种策略从拟南芥后代中获得重组系(transgression)非常有效(Van Berloo et al., 1999)。

分子标记辅助选择在 QTL 聚合选择中的优势显而易见,不足之处在于受可调控 QTL 数量的限制(Lecomte et al., 2004b; Gurand Zamir, 2015; Sacco et al., 2013)。这一点可以通过 QTL 的精细定位,或在其他背景下对该 QTL 的效应进行鉴定(Lecomte et al., 2004a)。目前 SNP 的利用和基因组选择为数量性状的标记辅助选择开辟了一条新途径。

3.2.5 番茄抗病虫育种

在番茄育种中,抗病虫性一直是优先考虑的因素。尽管长期以来不断通过表型选择和传统育种途径进行抗性改良,但在世界各地仍然有大量病虫害严重危害番茄生产。现代番茄遗传学和基因组学等新技术与传统育种方法相结合,通过聚合抗性基因或位点加速了抗性育种进程。

在番茄 MAS 抗性育种中所使用的标记包括:与抗性连锁的表型标记,例如对杀虫剂倍硫磷的敏感性与番茄细菌性斑点病抗性连锁(Laterrot et al., 1989);与抗性连锁的酶标记,例如 Aps-1^1 与根结线虫抗性连锁(Aarts et al., 1991, Messeguer et al., 1991)以及与抗性位点紧密连锁的 DNA 标记。通过这些标记将抗性位点聚合于番茄新品种中。MAS 的应用显著提升了选择效率,尤其是当某种病害的抗病性鉴定难度较大时,MAS 的优势尤为突出。分子标记可以提升目标位点导入的准确性和有效性,降低连锁累赘的负效应。MAS 选择能够将几个抗性位点与其他理想性状聚合到一起。在番茄基因组中,大多数抗性基因都是成串排列的,通过表型选择或利用主效抗性基因两边的标记进行 MAS 选择,就可以将这类两两连锁的抗性等位基因一并导入,形成一个具有多抗性的品种。例如,大多数携带 Tm-2^2 的番茄品种都携带着 Frl 基因,Frl 基因是针对番茄颈腐根腐病菌引起的根腐和冠状镰刀菌的抗性基因

(Foolad et al.，2012)。相反，如果抗性基因的连锁呈互相排斥状态，那么在育种中，就很难选择到纯合呈互引状态的重组系。例如，要将抗斑萎病毒的 *Sw-5* 和抗晚疫病的 *Ph-3* 结合到一起就很困难（Robbins et al.，2010）。总之，虽然 MAS 的应用显著提升了番茄改良的进程和效率，但是目前依然面临很多挑战。

目前，番茄育种中已经有 30 个非常重要的抗性单基因，这些基因都有相应的 DNA 分子标记可以利用（Foolad et al.，2012）。多基因复合遗传的抗性性状的 DNA 标记很少，几乎没有应用。MAS 已经作为一种常规方法用于对主效抗性基因的选择，如 *I*、*I-2*、*morer I-3*、*Ve*、*Mi-1.1/Mi1.2*、*Asc*、*Sm*、*Pto*、*Tm-2²*、*Sw-5* 等，因此，很多商业品种都具有对尖孢镰刀菌、黄萎病菌、根结线虫、链格孢菌、番茄细菌性斑点病、番茄花叶病毒和番茄斑萎病毒的抗性。同时 *Rx-3*、*Rx-4*、*Ty-1*、*Ty-2*、*Ty-3* 和 *Ty-4* 的分子标记也越来越多地用于对黄单胞菌和番茄黄化曲叶病毒抗性品种的培育中。

尽管分子标记已经广泛用于番茄抗病基因的鉴定，但并不是所有的标记都是有效标记。当种间杂交获得的育种群体中多态性缺乏、分子标记与基因距离太远或目标 QTL 与不良性状发生了互换，标记选择的可靠性会显著下降。然而，新一代测序技术可以鉴定开发 SNP 标记。这种标记是一种新的基于 PCR 的连锁标记，通过在育种群体中与目标性状的关联性开发获得。植物全基因组技术有助于在野生资源中利用 ecoTILLING、等位基因挖掘或利用全基因组关联分析技术鉴定与抗性性状连锁的、有用的分子标记。番茄育种家如今可以利用这些标记筛选出适宜的基因型组合，通过杂交获得相关理想性状，并设计培育理想的番茄品种。

3.3 番茄基因组选择育种

番茄大多数性状都是受多个微效 QTLs 调控，而连锁作图和全基因组关联分析存在一定局限性，表现为对微效 QTLs、稀有 QTLs 以及对环境条件高度敏感的关联性鉴定和量化分析方面（Crossa et al.，2017）。为了弥补这些不足，2001 年 Meuwissen 等人提出了基因组选择法（GS）（Crossa et al.，2017）：首先构建一个训练群体，这个群体具有覆盖整个基因组的标记（比如 SNPs）和表型数据；然后利用训练群体中的标记和表型信息对测试群体中未知表型个体进行遗传育种估值（genetic estimated breeding values，GEBV）。与表型选择相比，基因组选择法的主要优势在于能降低成本、节约时间

(Crossa et al., 2017)。

基因组预测（genomic prediction，GP）的准确性受以下几个因素影响：训练群体的大小、结构和遗传多样性，以及性状的遗传力、分子标记的数量和分布、连锁不平衡、预测方法、QTL 的数量等（Isidro et al., 2015；Spindel et al., 2015；Duangjit et al., 2016；Kooke et al., 2016；Yamamoto et al., 2016；Boison et al., 2017；Crossa et al., 2017；Minamikawa et al., 2017；Müller et al., 2017；Yamamoto et al., 2017；Crain et al., 2018；Edwards et al., 2019；Mangin et al., 2019；Sun et al., 2019）。为了提高预测的准确性，研究者研发了复合基因组选择模型。该模型考虑了不同因素的影响作用，如多性状和多环境的互作（Montesinos-López et al., 2016；Fernandes et al., 2018）。截至目前，已经开发出多个基因组选择模型，预测的准确性因性状和环境条件而不同（Heslot et al., 2012；Jonas et al., 2013；Yamamoto et al., 2016，2017）。

首次在番茄上使用基因组选择法进行设计育种和表型预测是为了达到番茄产量和口感同时提高的目的（Yamamoto et al., 2016），这同时实现了基因组选择法从一个理论方法到实际育种中的运用。通过这种方法测定了 20 个广义遗传力在 0.10~1.00 之间的农艺性状，筛选出 96 个大果番茄品种，并根据果实产量、品质和生理性异常等指标将其划分为 4 组。基因组预测模型有 7 个，其中 5 个为线性模型，包括岭回归法（ridge regression，RR）（Endelman，2011）、贝叶斯套索法（Bayesian lasso，BL）（Park et al., 2008）、拓展的贝叶斯套索法（extended Bayesian lasso，EBL）（Mutshinda et al., 2010）、加权贝叶斯收缩回归法（weighted Bayesian shrinkage regression，wBSR）（Hayashi et al., 2010）和贝叶斯 C 法（Bayes C）（Habier et al., 2011）；其余 2 种为非线性模型，即再生核希尔伯特空间回归法（reproducing kernel Hilbert space regression，RKHS）（Gianola et al., 2008）和随机森林法（random forest，RF）（Breiman，2001）。不同模型对不同性状预测的准确性不同。其中贝叶斯 C 法对 8 个性状的预测准确度都最高，位居 7 个模型之首。筛选果实总重和可溶性固形物含量遗传育种估值（GEBV）高的单株作为亲本进行后代模拟，经过 5 个世代的模拟结果表明，模拟的 GEBV 与亲本相当。育种中，对目标性状进行选择的同时也会对一些非目标性状产生影响。例如在对产量和风味进行选择时会引起植株形态变化，如株高增加。以上研究结果表明，事先利用模型对实际育种方案进行模拟很有益处。

Yamamoto 等人（2017）利用大果番茄 F_1 群体构建基因组选择模型，用于预测在实际育种中果实总重和可溶性固形物含量的提升潜力。6种基因组选择模型测试和10倍交叉验证表明,对可溶性固形物含量预测的准确度高于果实总重。就可溶性固形物含量预测而言,GBLUP模型和贝叶斯套索法预测的准确度显著高于其他模型;对于果实总重的预测准确度,再生核希尔伯特空间回归法和随机森林法模型显著高于其他线性模型。Yamamoto等人进一步预测4个后代群体的性状分离情况,结果表明4个后代群体中所有的单株在遗传上彼此不同,而表型值均居于双亲之间。每个群体的遗传多样性均明显低于训练群体。

Duangjit等人(2016)研究了影响基因组预测效率的几个关键因子,包括训练群体的大小、SNP的数量和密度,以及单株个体之间的关联度。通过对163个番茄材料的分析,他们认为训练群体的最适样本大小为122个,预测的准确度随标记数量和密度的增加而提升,但是提升幅度很小。个体之间的关联性也影响预测的准确度,个体之间的关系越紧密,预测的准确度越好。然而,这个研究中也存在一些局限性:①仅仅测试了岭回归最佳线性无偏差预测统计模型(Endelman,2011);②SNP数量相对较少,而且对基因组中某些区段的覆盖度十分有限(Zhao et al.,2019);③群体的结构中,野生材料的数量占比显著低于樱桃番茄和大果番茄。

大多数基因组选择模型都建立在标记信息之上,标记之间存在的上位性互作很难被发现。通过连锁和连锁不平衡分析,可以将分子标记组合到单倍型中,以提高预测的准确度(Clark,2004;Calus et al.,2008;Jiang et al.,2018),关于这一点最近已经在动物中得到验证(Calus et al.,2008;Cuyabano et al.,2014,2015a,2014b;Hess et al.,2017;Karimi et al.,2018)。这种基于单倍型的基因组预测模型可以挖掘存在于单倍型中的局部上位效应(Wang et al.,2012;De Los Campos et al.,2013;He et al.,2016;Jiang et al.,2018)。关于单倍型基因组选择在主要作物中的优势还需要进一步发掘(Jiang et al.,2018)。

采用基因组选择能够在育种中将多个性状结合起来同时进行选择;能够在考虑产量和品质相关性状遗传架构的同时,将对生物胁迫的质量型以及数量型抗性性状和对生物和非生物胁迫耐受性组合起来;将前景选择和背景选择结合起来能够促进在各种环境变化条件下产生持续稳定的表现。截至目前,病害的数量型抗性还没有被育种家所推崇和利用,主要是因为多基因调控的复杂性阻碍了QTL的广泛使用。后基因组学的发展将有助于加速番茄包括多种病虫抗性性状在内的多基因性状育种的发展。

4 番茄理想型的模型设计

20世纪70年代,番茄育种就开始趋向于培育能够适应机械化、高投入生产体系的高产品种。90年代以来,环境、经济和社会诸方面的变化促使育种目标不断更新。育种不仅需要遗传学、生态生理学和农艺学等多学科、多领域结合进行,还要考虑表型和基因型的连锁机制,环境特别是土壤、气候和虫害对番茄影响作用的模式以及栽培管理技术体系和水平。这就要求在当前生产体系下,对基因型和环境互作、植株的产量和品质表型以及环境的影响作用进行有效评估,也需要将通过遗传和基因组学工具获得的遗传信息与决定目标农艺性状变化的表型性状结合起来。在这种背景下,理想型的概念逐渐形成,即按照设定的生产体系去设计与之相适应的植株表型,再按照这些表型制定最终育种目标。为了这个目的,学界已经开发了一些基于过程的预测模型,这些模型已经应用于揭示复杂性状的遗传变异机制(Reymond et al.,2003;Tardieu,2003;Quilot et al.,2005;Struik et al.,2005)、分析基因型与环境与管理三者之间的相互作用(Génard et al.,2010;Bertin et al.,2010;Martre et al.,2011),以及设计能够适应特殊环境的新的理想型(Kropff et al.,1995;Quilot-Turion et al.,2016;Martre et al.,2015;Génard et al.,2016),这些模型的有效性也在应用中得到了证明。

4.1 番茄理想型的概念

理想型就是一种理论上的生物学模型,这种模型能够在给定的环境中按照预先设定的方式运行(Donald,1968)。理想型的概念最先在小麦上应用,之后被拓展到其他几种作物上。2015年,Martre等人将原有理想型的定义拓展为将形态性状和生理性状或他们的遗传基础结合到一种作物中,使这种作物能够很好地适应特定的生物生理环境、作物管理方式以及最终使用的方式。

单基因性状如生物胁迫抗性在育种中可以被直接利用。例如,Zsögöna等人(2017)提出基因组编辑技术在性状改良方面具有优势,一方面可以对栽培品种中的单基因性状进行编辑,另一方面也可以对含有多基因胁迫抗性的野生种中的产量相关性状进行编辑。实际上，多基因控制性状的改良要复杂得多，这也是遗传学家要面对的主要难题。具体表现为：①选择目标性状的复杂性，如产量、品质、氮利用率、对水分胁迫的适应性，这些性状受到很多基因的调控，是多种具有反馈效应的呈网状结构的生物过程作用的结果；②这些性状的表达均受到环境和栽培管理技术的影响。这些复杂性导致基因型、环境、管理三者的互作增强，进而增加了这些性状遗传研究和育种应用的难度。这种经验性方法的首次尝试是针对一种特定环境和生产系统，于是就设计出了与之相适应性状的最佳组合。而针对很多不同复杂情况时，基于过程的预测模型在理想型设计中也具有重要作用（Quilot-Turion et al.，2012；Génard et al.，2016）。

4.2　番茄基因型、环境、管理互作（G×E×M）的预测模型

植株及其器官可以看作是一个复杂的系统，该系统在基因型、环境、管理三者的相互作用（G×E×M）下，系统内各种生物过程在不同层次和规模上交互作用。基于过程的预测模型是对这个系统进行的数学描述，通过对不同组织器官内所有生物过程的整合，实现从细胞到整个植株所有过程的整合，以模拟植株在与环境互作中的复杂性。在预测模型中，所谓的构成性状都作为复杂性状，用模型参数来表示。用参数代替复杂性状本身，可以将模型参数与相关的遗传变异直接关联起来（Struik et al.，2005；Bertin et al.，2010）。模型参数的确定通常涉及正向遗传学研究方法，如QTL作图法，一方面用于寻找性状相关QTL之间的共定位，另一方面可以找出用于模型参数的QTL（Yin et al.，1999；Reymond et al.，2003；Quilot et al.，2005；Prudent et al.，2011；Constantinescu et al.，2016）。因此，建模要先鉴定模型参数，找出模型中依赖基因型变异的特异性参数以及与之相反的不随基因型变化的其他遗传参数，然后将每种基因或等位基因的组合用一组参数来表示，最后就可以对不同环境和栽培管理条件下的表型进行生物信息学水平上的模拟。要将预测的范围从已知基因型拓展到未知基因型，就必须对涉及的基因型的参数值进行估算。基因型参数值的估算需要将基于建模的QTL或QTL模型、等位基因、基因或基因模型组合起来进行（Martre et al.，2015）。如果将每个独立性状看作是基因型和环境效应的组合体，那么相较于传统QTL作图，基于模型的研究途径能够检测出更多更稳定的QTLs。然而，截至目前，仅有一小部分基

因型参数如等位基因变异能用于番茄模拟模型的构建（Prudent et al., 2011; Constantinescu et al., 2016）。

目前已经构建了几种基于过程的模拟模型，用于对果实发育和品质形成过程进行预测。利用这些模型可探索不同基因型、环境、管理三者组合互作的奥秘（Génard et al., 2004; Bertin et al., 2010; Martre et al., 2011; Kromdijk et al., 2013）。番茄上也建立了几种植株模型，这些模型都是根据不同优先原则，由碳汇中、碳同化和碳分配的过程构成（Heuvelink et al., 1994; Jones et al., 1991; Boote, 2016; Fanwoua et al., 2013）。同时，番茄上模拟水分转运和积累的模型也有几个。例如，Lee (1990) 建立了番茄果实单位面积水分吸收和蒸发的单方向恒定流量模型；Bussières (1994) 基于水势梯度和阻力，建立了番茄果实的水分输入模型。但将干物质和水分积累结合在一起的果实发育模型目前还未有研究。梨果实模拟模型已经用于对番茄果实发育和组成的过程预测（Liu et al., 2007; Fishman et al., 1998）。这个模型以发生在一个细胞中的所有生理过程为基础，其中糖分的运输是通过果实韧皮部以质量流、扩散和主动运输的方式进行，水分流主要由细胞壁内外水势差调控，而且只有内外水势差足够大时才能有效发挥作用。研究人员已经陆续对这些模型进行了改良和完善，将其与预测木质部和韧皮部在运输方面的作用和贡献的茎发育模型相结合（Hanssens et al., 2015），用于评估作物负荷量对果实发育的影响（De Swaef et al., 2014）。

将这种果实模拟模型与植物结构模型相结合，可用于预测植物体内水分和碳素的分配与分布以及由碳素引起的水势梯度和韧皮部汁液的浓度梯度（Baldazzi et al., 2013）。由于细胞是果实代谢模型的基本单元（Génard et al., 2010），因此建立细胞分裂和体积增大的发育过程模型是构建果实发育模型的关键（Baldazzi et al., 2012; Okello et al., 2015）。目前还没有将细胞分裂、细胞扩增和DNA核内复制都包括在内的番茄果实发育模型。如果一个果实模拟模型能够预测细胞发育过程之间的相互作用，那么这个模型就能够将亚细胞模型整合在一起（Beauvoit et al., 2018）。例如，有学者提出了番茄果实发育模型，用这种模型可以描述果实发育过程中代谢物质的转化（Colombié et al., 2015, 2017），并可根据酶活性及其区域分布来推测果皮可溶性糖的含量（Beauvoit et al., 2014）。实际上，目前除了糖代谢以外（Prudent et al., 2011），尚未有预测果实其他物质构成的模型，这是果实品质研究的一个主要课题。例如，缺少对果实主要营养保健物质如类胡萝卜素、维生素的预测模型。目前已经构建的梨酸度预测模型（Lobit et al., 2003,

2006）或许可以用于番茄果实中。

这种以果实为中心的综合模型，通过整合细胞水平的生物过程并连接形成一个完整的植株模型，开启了将果实发育和物质构成的分子调控信息结合在一起用于分析基因型、环境、管理三者互作对产量和品质的影响作用（Martre et al.，2011）。事实上，综合模型是用生物信息学分析植物表型的重要工具。综合模型不仅可以预测植株和器官性状，如产量或果实组分，还可以估算一些不易大规模测量的生理学变量，如木质部和韧皮部质流、糖分主动转移等（Génard et al.，2010）。因此，基于过程的模型有助于更好地理解遗传变异和鉴定候选基因，帮助育种家鉴定最相关的性状并对最适宜发育时期进行表型分型，从而在给定环境参数的条件下，研究基因型和表型之间的关系（Struik et al.，2005）。

4.3 番茄理想型的模型设计

如何模拟基因型、环境、管理三者互作关系是理想型生物信息学设计的关键，即如何在特定条件下通过多元优化法将与果实发育和品质相关的 QTL、基因、等位基因组合起来，以实现果实生长和品质发育的最优化（Quilot-Turion et al.，2016）。这里就有一个利用过程预测模型进行育种方案制定的问题。

一个基于过程模型的育种计划可分为三个连续的步骤，如图 4-1 所示。第一步，确定目标性状。在给定的生产环境条件下，如低水分供应、植株整枝方式等，确定构建模型所需要的遗传参数值，根据这些遗传参数值确定实现理想型或模拟表型所需要的性状。第二步，确定目标性状相关基因的综合效应值。从遗传学角度，通过模拟基因型估算这些遗传参数值，并给出与每个遗传参数相关的等位基因的组合效应值。第三步，确定理想基因型。可以在已有的基因型中找出与给定环境条件下设计的理想基因型关联最紧密的基因型，或者提出育种策略以获得符合理想型要求的新的基因型。在最后一步中，过程模型可以与遗传模型相结合，模拟在育种过程中期望出现的基因型变化。Quilot-Turion 等人（2016）考虑到基因多效性和连锁效应，提出可通过增加遗传约束或限制条件，直接优化调控参数的等位基因以促进理想基因型的实现。这种途径能够重新建立起由现实后代获得的各种参数之间的相互关系，利用这种关系，可以在设定环境条件下，找出改进果实表型的等位基因的最佳组合。

尽管过程模型具有明显的优势和应用前景，但截至目前，利用该模型设计

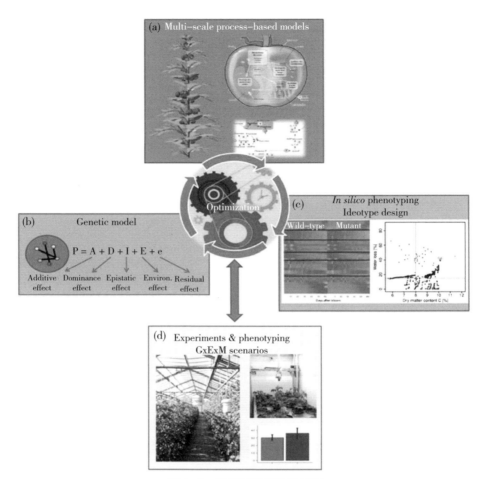

图 4-1　番茄理想型设计过程

在可控环境下对植物或器官表型进行测量，或在不同基因型、环境、管理组合作用下的表型分型，平台 d 可通过植物中水分和碳流描述、生长过程描述、果实初生和次生代谢描述的耦合过程模型进行预测。图 c 说明如何将植物和果实表型分型模型结合起来进行理想型设计。热图表示，当模型中一个遗传参数虚拟突变时对所有模拟过程的影响作用，方块表示根据模型得出的理想型的位置，模型是在水分缺乏条件下根据果实干物质含量和果实水分流失情况建立的。图 b 多效应遗传模型，控制着图 a 过程模型中的基因型参数。遗传模型能够预测图 c 中筛选出的理想型所对应的基因型。优化过程可以同时用于预测模型中的基因型参数和设计理想型。

出来的番茄理想型还很少。例如，Sarlikioti 等人（2011）利用静态功能结构植物模型寻找适宜温室栽培且有利于光吸收和光合作用的番茄植株最适结构。研究认为，具有长节间和长而窄叶片结构的植株其光合作用能力强。又如，根据上面介绍的番茄果实模拟模型，Constantinescu 等人（2016）提出了一个在缺水条件下通过将糖分的高主动吸收和果柄的高导度结合起来以实现保持大果基因型番茄产量和品质的有效策略。通过模型标定，研究人员在其研究群体中鉴定出了一些与理想型很接近的基因型，这些基因型所携带的有益性状和等位基因将用于适应低水分供应的植物品种的选育。

如上所述，人们期望用于理想型设计的预测模型机械化程度要高，涉及的QTLs或基因等参数要更具体，这就需要对描述参数进行不同层次和范围的组合。因此，预测模型的建立过程非常复杂。模型参数要通过足够数量的表型分析获得理想的估量，或是通过对模型校定进行更准确的估算获得。表型的参数化是最难的，也是限制性因子，需要建立在大量的大规模遗传试验的基础上（Cournède et al.，2013）。同样，在模型校定这个环节，对QTL、等位基因或基因模型参数的预测也会受到参数化基因型数量有限的困扰（Letort et al.，2008；Migault et al.，2017）。要取代对研究群体所有基因型大量生理性状的测量工作，可以选择能够很好地反映该群体遗传多样性的一组基因型，然后通过QTL或是基因组预测模型去预测所有入选基因型的参数（Van Eeuwijk Fred et al.，2019）。还有一种做法由Constantinescu等人（2016）实践得出，即根据相关形态及生理性状筛选出一套参照基因型用于估算模型参数。从数学的观点看，理想型设计是一个要依靠多目标优化法的复杂过程，其复杂性一方面在于维度和量纲的问题，如基因型和变量的数量不断增加，另一方面在于理想型通常是一些非线性甚至是对抗性性状的结合，番茄果实的产量和品质性状就是这样的例子。为了解决优化问题，可采用以不同算法为基础建立的元启发式算法的大模块，能够在合理时间范围内给出满意方案（Ould-Sidi et al.，2011）。这些方法同样适用于模型的标定。

随着高通量基因分型和表型分型平台的出现及大批量高时空维度的植物形态和生理数据库的产生，人们处理大规模大批量表型的能力显著提高。表型分型信息与不同的基因型—表型模型的有益结合，已经在大田作物中得到验证（Van Eeuwijk Fred et al.，2019）。然而，在番茄和其他园艺作物中，平台中表型性状的范围还需要在日常观察性状的基础上进一步扩大。如除了需要植株和果实发育性状之外，还需要包括果实生长和组分性状随植株和果实发育的变化情况。

4.4　番茄模型设计育种的应用前景

植物模型设计无论是在作物管理还是在针对不同环境和多个育种目标的植物育种中均具有广阔的应用前景，番茄研究中尤为如此。番茄基因组序列已知、遗传资源丰富、已经在不同器官水平上建立起了包含不同过程网络的过程模型，以及消费者对高品质果实的强烈需求等，都是支撑番茄理想型成功设计的关键要素。细胞和分子水平的整合有助于植物模型的重塑，有助于揭示不同时空互作对目标性状调控的复杂性。为此，针对参与模型过程的基因，建立其相应的工作小网络，将有助于把过程模型参数与其遗传基础联系起来。

尽管概念是合理的，但截至目前，几乎没有一个植物改良是根据生物信息学的理想型设计来进行的。因此，模型学家、农学家、遗传学家和育种学家应加强协作，将不同学科的优势结合起来。对番茄而言，需要将过程模型和遗传模型结合起来，需要进一步开发新的亚过程模型，用于预测番茄重要品质性状如质地、类胡萝卜素含量和维生素含量等。

我们要对通过基因聚合进行遗传改良这样一个主流做法提出质疑。事实上，将多个基因聚合到一个品种中的确可以有效提高作物对生物胁迫的多抗性，但对于那些依赖基因数量、遗传结构、资源特性等的性状，这种做法就很难奏效（Kumar et al.，2016）。取而代之的一个新观点就是如何在降低投入的同时使多基因型作物的表型稳定。这需要我们对于一个群体基因组之间的相互作用有更深刻、更正确的理解。

5 番茄生物技术与基因工程

5.1 番茄基因工程技术的发展

根据 2017 年国际农业生物技术应用公司年报，全球有 24 个国家 1.7 亿农民在种植生物技术或转基因作物，面积达 1.898 亿 hm^2。尽管目前市场上没有基因工程番茄，但是却有名为 FLAVR SAVR 的番茄品种——第一个通过基因工程育种并作为商业化食品供应市场的番茄品种。该品种于 1994 年 5 月 18 日获得美国农业部食品药品监督管理局的许可。这个品种是由美国加尔基因公司的科学家通过对多聚半乳糖醛酸酶进行 RNA 的反义调控获得的。PG 是含量最丰富的蛋白之一，长期以来，人们认为其主要作用是负责成熟番茄的软化 (Kramer et al., 1994)。在 FLAVR SAVR 品种中，PG 含量降低了 99%，番茄果实储存过程中的软化现象显著减少，成熟果实对真菌侵染的抗性明显增强，进而延长了果实的货架期。科学家期望基因工程番茄能够在藤蔓上自然成熟以提升风味，同时还能够适应传统的配送集散系统 (Kramer et al., 1992)。就在同一年，捷利康种子公司制作了一种商业化的番茄酱，所用的番茄果实来自通过正义基因沉默多聚半乳糖醛酸酶的番茄品种，其果实的黏度和风味有所提升，损耗下降 (Grierson, 2016)。1999 年 FLAVR SAVR 退出市场。截至 1999 年，获注册的基因工程产品有很多，但都没有商业化（表 5-1）。2000 年以来，还没有新的转基因番茄注册。

表 5-1 获准商业化的转基因番茄品种

品种	发明者	性状	年份	获批机构	国家
FLAVR SAVR	加尔基因公司	延迟软化	1994	在美国全面使用，在日本、墨西哥用作饲料	美国
1345-4	DNA 植物技术公司	延迟成熟（删减氨基环丙烷环化合酶基因）	1994	在美国全面使用，在加拿大和墨西哥用作食品	美国
Da,V,F 番茄	捷利康种子公司	延迟成熟	1994	在美国全面使用，在加拿大和墨西哥用作食品	美国

(续)

品种	发明者	性状	年份	获批机构	国家
8338	孟山都公司	延迟成熟[accd 基因（1-氨基环丙烷-1-羧酸脱氨酶）]	1995	在美国全面使用	美国
351 N	Agritope	延迟成熟（SAMK 基因）	1995	在美国全面使用	美国
Huafan No 1	华中农业大学	延迟成熟（反义 EFE 基因）	1996	没有数据	中国
5345	孟山都公司	抗虫（引入 cry1Ac 基因）	1997	在美国全面使用，在加拿大用作食品	美国
PK-TM8805R（8805R）	北京大学	延迟成熟	1999	在中国种植，用作食品、饲料	中国

5.2 番茄基因工程技术

番茄的遗传转化技术始于 20 世纪 80 年代（McCormick et al.，1986）。农杆菌介导法是基本的遗传转化模式，即将番茄的叶片、下胚轴或子叶作为外植体与农杆菌进行共培养，通过愈伤组织诱导成芽，芽进一步形成再生植株。根据 Bhatia 等人（2004）总结的操作指南，番茄基因工程一般程序如图 5-1 所示。

具体步骤如下：
①载体将基因工程目的基因表达单元转给农杆菌，由农杆菌再转给植物。
②转入的基因工程目的基因表达单元与植物基因组整合实现稳定转化。
③转化植株的离体再生和筛选。

有效转化和再生是利用基因工程的前提。转化效率很大程度上取决于基因型、外植体以及培养基中植物生长调节剂的种类和含量（Gerszberg et al.，2015）。

利用农杆菌悬浮液蘸花或注射花蕾也可以成功地进行遗传转化。Yasmeen 等人（2009）利用这种方法获得了较高的转化效率，其效率在 12%～23%；而 Sharada 等人（2017）利用蘸花或注射法得到的转化效率却很低，其效率在 0.25%～0.50%。在拟南芥中，蘸花法已经成为广泛使用的遗传转化途径（Clough et al.，1998）。但这种方法在番茄上并不是很有效。

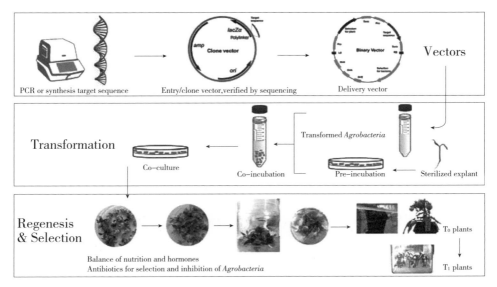

图 5-1　番茄转基因流程图

目的基因序列可通过 PCR 获得或由商业合成，然后采用不同的克隆方法将其转移到克隆载体上。克隆载体经过验证后，将目的基因序列由克隆载体转移到运载载体，该载体用于农杆菌转化。番茄种子在无菌培养基上发芽，待子叶出现后剪切进行预培养，接着与携带运载载体的农杆菌共同培养 2 d，然后将外植体转移到筛选培养基上进行诱导培养。不同阶段采用不同的激素和营养配方进行培养。T_0 代植株需要鉴定确认目的序列已经插入，再将 T_0 代植株收获的种子播种在含有筛选抗体的培养基上进行转基因植株筛选。

番茄中利用异源基因的沉默或过表达已经有几十年历史了，而在番茄上使用基于成簇的规律间隔的短回文重复序列（CRISPR/Cas9）的基因组编辑技术仅有几年时间（Brooks et al., 2014）。与传统基因工程方法不同，CRISPR/Cas9 技术很快就表现出在基因功能鉴定、育种和驯化等方面的巨大潜力。

5.2.1　番茄基因沉默与过表达技术

基因沉默通常是通过基因的反义、正义或 RNA 干扰来实现的。这项技术通常用来延缓番茄果实的成熟过程，抑制采收后果实的软化，延长果实货架期以及便于长距离运输，也用于去除番茄果实中的过敏原类物质（Le et al., 2006）及阻止种子的形成以获得单性结实的果实（Schijlen et al., 2007）。抑制或更好地控制果实的成熟和软化至今仍然是困扰育种家和番茄商业化前景的主要挑战之一。为了实现这一目标，研究者曾对不同基因进行了不同程度的沉默，包括编码果胶甲酯酶基因（Tieman et al., 1994）、延展蛋白基因

（Brummell et al.，1999）、β-半乳糖苷酶基因（Smith et al.，2002）、ACC 合酶基因（Gupta et al.，2013）、SlNAC1 转录因子（Meng et al.，2016）以及果胶裂解酶基因（Uluisik et al.，2016）。

基因沉默技术主要用于使番茄的内源基因下调表达。而基因过表达技术既可以针对内源基因也可以针对外源基因，常用于启动子和基因的表达、增强对生物和非生物胁迫的耐受性、提高次生代谢物质的积累等方面的研究。内源或外源的启动子都可以与 GUS 或荧光蛋白基因融合，用来跟踪研究基因的表达模式。Fernandez 等人（2009）根据果实发育和成熟过程中系列启动子的表达模型和特性设计了 Gate-way 靶向载体，加速了番茄基因工程的发展。氧化还原感应器 GFP 的建立可以更好地研究番茄活体的氧化还原状态（Huang et al.，2014）。

研究苹果、桃、香蕉等多年生植物的人员通常利用番茄作为模式植物，通过异源活体表达研究目标基因的功能。一方面，这些物种的遗传转化和再生技术难度较大；另一方面，即使技术允许，这些物种从幼苗到获得果实表型需要的时间也很长。反过来，如果其他物种的基因在番茄上有表型，那么这个基因也可作为番茄基因工程可以利用的基因资源。如在番茄中过表达苹果液泡中的 H^+ 转运无机焦磷酸化酶（MdVHP1），能提高番茄对盐胁迫和水分胁迫（干旱）的耐受性（Dong et al.，2011）；在番茄中过表达香蕉的转录因子（*MaMYB3*），可抑制番茄果实中淀粉降解，延迟果实成熟（Fan et al.，2018）。

将非生物驱动的启动子与响应非生物胁迫的转录因子融合在一起进行遗传转化，可以有效提高番茄的胁迫耐受性。含有由 ABA-响应复合体（*ABTC1*）驱动的转录因子 *CBF* 的转基因植株在正常生长条件下，对冷害、水分胁迫（干旱）和盐胁迫的耐受性增加，植株的生长和果实产量却不受影响（Lee et al.，2003）。

代谢通量也可用于改良番茄果实品质，如调控挥发性物质和营养物质。Domínguez 等人（2010）通过过表达编码 ω-3 脂肪酸去饱和酶、*FAD3* 基因和 *FAD7* 基因，使转基因番茄叶片和果实中的顺-3-己烯醛与己醛的比例显著增加至 18∶3 和 18∶2。果实特异性启动子 *E8* 与 *AtMYB12* 融合插入番茄基因组，激活了与黄酮醇和羟基肉桂酸酯生物合成相关基因，使其积累量增加到果实干重的 10%（Zhang et al.，2015a，2015b）。

20 世纪 80 年代以来，基因工程取得了很多标志性的成果。随着基因编辑技术如 CRISPR/Cas9 的出现和发展，更具影响力的成绩已经陆续出现。

5.2.2 番茄基因组编辑技术

与基于蛋白-DNA 识别的锌指核酸酶（ZFN）和转录激活效应样因子核酸酶（TALEN）等早期基因组编辑工具不同，CRISPR/Cas9 依赖于简单的 RNA-DNA 碱基配对和前间区序列邻近基序（PAM）序列识别（Gaj et al.，2013）。所有基因组编辑工具的共同点在于引起 DNA 双链裂解（DSBs），但是 CRISPR/Cas9 的效率要比 ZFN 和 TALEN 高得多（Adli，2018）。DNA 双链裂解的修复方式有两种，一是通过易错非同源末端连接（NHEJ），二是通过同源定向修复（HDR）。生物自身通过 NHEJ 或 HDR 修复系统，诱导缺失、插入突变或精准置换，实现基因敲除或精准编辑。在研究 CRISPR/Cas9 基因组编辑机制的同时，科学家们对 CRISPR/Cas9 工具的再次工程化表现出极大热情，希望该系统使用起来更便利、更精确，使 Cas9 核酸酶变得更小、靶点范围变得更大、脱靶的频次降到更低。

2014 年，首例 CRISPR/Cas9 在番茄上应用（Brooks et al.，2014），接着，基于 CRISPR 的工程技术相继应用到多个领域。CRISPR/Cas9 系统能够有效诱导或敲除突变，是鉴定自然突变或正向遗传学中候选基因的一种有效方法。最典型的案例就是利用 CRISPR/Cas9 技术诱导形成 *RIN* 敲除突变体。番茄 *rin* 突变体的果实不能形成乙烯和红色色素，在采收后仍然保持坚硬，因此人们长期以来都认为 *RIN* 敲除突变体对于诱导果实成熟是必不可少的。但利用 CRISPR/Cas9 诱导的 *RIN* 敲除突变体破解了这个观点。Ito 等人（2017）通过 CRISPR/Cas9 基因编辑技术获得了 *RIN* 敲除突变体，该突变体的果实呈现出与 *rin* 果实不同的表型：*rin* 果实完全不能成熟，而 *RIN* 敲除突变体敲除突变体果实呈现微红色。此外，用 CRISPR/Cas9 编辑 *rin* 突变体等位基因也部分恢复了诱导成熟的表型。这些结果表明，*RIN* 敲除突变体并不是成熟起始的必需因子，而是一个功能获得型突变，并非无效突变，其产生的蛋白能够有效抑制成熟。这项技术还可用于甲基化或去甲基化的研究中。番茄中的一个 DNA 去甲基化酶基因 *SlDML2*，就是通过 CRISPR/Cas9 诱导产生的一个功能缺失型突变体，该基因对番茄果实成熟具有决定性作用，其作用可能是通过去甲基化作用激活成熟诱导基因，同时抑制成熟抑制基因（Lang et al.，2017）。

二代 CRISPR 基因编辑工具包括单碱基编辑、CRISPR 介导的基因表达调控和 CRISPR 介导的活体细胞染色质成像（Adli，2018）。随着着陆平台（Danilo et al.，2018）的产生以及精准碱基突变的基因插入的实现（Danilo et al.，2019；Veillet et al.，2019），基因插入的可能性得到了进一步提高。所有这些都建立在对 Cas9 改造的基础上，原因在于即使将 Cas9 核酸酶转化为 Cas9 核酸内切酶（nCas9，nickase Cas9）或灭活 Cas9（dCas9，催化失活 Cas9），dCas9 仍然具有识别特异序列的能力。经过工程化改造的 Cas9 可以与其他酶或蛋白质融合，从而实现碱基编辑、基因调控或染色质成像。

Shimatani 等人（2017）将 D10A 突变体 *nCas9At* 与人类优化了的密码子 *PmCDA1*（nCas9At-PmCDA1Hs）或拟南芥优化了的密码子（nCas9At-PmCDA1At）融合，创制了无选择标记的纯合可遗传的 DNA 置换植株。值得一提的是，由 T_0 代产生的后代仍然表现出插入缺失，而置换率与插入缺失突变相比要低得多，说明通过碱基编辑进行作物改良的可行性还比较低。Dreissig 等人（2017）通过 eGFP/mRuby2 与 dCas9 融合以及 CRISPR-dCas9 与荧光标记蛋白的结合，引起生物体内 DNA-蛋白互作。这种作用可以体现为烟草活体叶片细胞中染色体端粒的重复。此外，研究人员还开发了 CRISPR 的干扰（CRISPRi）途径，利用 dCas9 的绑定活性阻断转录过程以实现基因的下调表达（Qi et al.，2013）。

CRISPR/Cas9 与二代基因组编辑技术相结合扩大了生物技术的应用范围，提高了生物技术应用的可行性。这些技术的进步与传统转基因手段（RNA 干扰、过表达等）相结合，足以支撑通过综合育种途径以应对人口增加和气候变化带来的多种挑战。

5.2.3 番茄综合基因组工程

Rodriguez-Leal 等人（2017）利用 CRISPR/Cas9 编辑技术，通过对启动子的基因组编辑获得了多个顺式调控等位基因，并利用这些等位基因改良了番茄的果实大小、花序分枝和植株结构共 3 个主要产量性状。通过评估这些等位基因变异的表型效应，可为筛选和固定控制数量性状的新等位基因提供有效途径。

有两个成功案例表明基因组编辑可以加速番茄驯化过程。Li 等人（2018）先筛选了 4 个对胁迫具有耐受性的野生番茄材料，采用非多重 CRISPR/Cas9 编辑技术，靶向编辑与形态学、花和果实发育、维生素 C 合成相关基因的编码序列以及顺式调控区或上游开放阅读框，将一些理想性状添加/插入/敲入这 4 个材料中。他们发现，经过编辑的植株后代既具有驯化后的表型，也保持着亲本的抗病性和耐盐性。Zsögön 等人（2018）以野生醋栗番茄为起始亲本，通过对 6 个产量和产能重要位点的编辑，实现了野生材料中有用性状与期望农艺性状的结合，获得的基因工程番茄的果实大小、数量和番茄红素含量均显著增加。正如研究人员描述的那样，这种令人惊叹的再次驯化案例为挖掘野生植物中的遗传多样性开辟了新的方向。

同时，基因组编辑工具也显示出在实现番茄理想型方面的巨大潜力。最近，Naves 等人（2019）提出利用基因工程将番茄改造成次生代谢物质的生物工厂，如辣椒素（引起辣椒灼烧感的次生代谢物质）。考虑到番茄基因组中含有辣椒素合成所需的所有基因，提出两种方案：一种是通过转录活化样效应因子（TALEs）替换靶定的启动子，另一种就是通过基因组工程替换靶定的启动子。这两种方式都可用于激活番茄中的辣椒素合成（Naves et al.，2019）。

5.3　番茄病虫抗性的基因工程

对于番茄的某些病害，至今还未发现针对它们的自然抗性基因或 QTL。尽管可以从野生近缘种中获得抗性基因，但由于种间杂交障碍或种间杂交常带来连锁累赘，这些抗性基因很难被育种者充分利用。在这种情况下，番茄的病虫抗性就可以通过生物技术手段进行改善。

为了弥补番茄自然抗性资源的缺乏，包括 RNA 干扰或过表达病原驱动序列等在内的转基因技术已被用于人工创制一些对病原菌具有抗性的种质。此外，其他作物抗性基因的异源表达也能用于增强番茄的抗性。如将来自中国辣椒（*Capsicum chinense*）的隐性基因 *pvr1* 渗入番茄，番茄就能获得对马铃薯 Y 病毒的显性广谱抗性（Kang et al.，2007）。Nekrasov 等人（2017）也通过基因组敲除创制了非转基因的抗白粉病番茄种质。

CRISPR/Cas9 编辑技术有望加速番茄抗病品种的选育。最近，CRISPR/Cas9 编辑技术已经用于创制对番茄黄化曲叶病毒免疫的工程番茄植株。这些植株是通过对番茄黄化曲叶病毒基因组中编码病毒外壳蛋白序列引导 RNA 的 Cas9 单敲除，或是对编码病毒复制酶基因的 Cas9 单敲除获得的（Tashkandi et al.，2018）。尽管该技术目前还处于初期阶段，但根据已有相关物种中抗性基因的进化知识，如 Bastet 等人（2019）介绍的拟南芥中相关基因，可以预见，基于 CRISPR-nCas9-胞嘧啶脱氨酶技术的基因编辑将用于番茄功能性抗性等位基因的设计和合成。

5.4　基因编辑植物的监管

2013 年以来，CRISPR/Cas9 编辑技术的应用使得植物基因组编辑取得了长足发展。与之前的技术相比，该技术成本低，转化效率高，迅速为广大研究人员所利用。然而，作为一项新兴的技术方法，该技术仍处在不断地发展和完善中，要充分认识和利用其全部潜力还需要科研人员不断探索。这项技术提供了难得的发展机会，同时也对如何监管提出了挑战。已有观点认为，由基因编辑获得的植株和通过传统遗传改良的生物体均为非转基因植株。很多植物育种家和科学家认为，由基因编辑如 CRISPR/Cas9 编辑技术获得的植株应归为突变，由于这种突变仅改变了 DNA 序列而不是外源基因的插入，因此应有别于转基因生物（GMO）。而持反对意见的人主张，人为通过基因编辑途径改变其本来的特性的作物都应该属于 GMO。在美国、加拿大及其他几个国家，CRISPR/Cas 编辑诱导的突变虽被认为等同于传统育种，但在法律上却有别于 GMO。2018 年 7 月 25 日，欧洲法院裁定基因编辑作物与传统 GMO 在法律上相同（Callaway，2018）。这可能对不同国家的育种发展产生较大的影响。

6 结论与展望

番茄是一种适应性很强的作物，可在世界各地广泛种植。相应地，种植环境的差异使番茄必须对来自不同环境的各种胁迫做出响应。分子标记的出现和使用可以将复杂性状的遗传基础解析分解为单个组分，进而找到和给出相关基因或 QTL 在染色体上的位置，最终使性状筛选工作变得更为便利和简单。另外，分子标记的使用也显著提高了育种者对野生种的利用效率和利用方式。由分子标记锚定的野生种基因组区域构成的异源文库为现代品种农艺性状的改良提供了一个重要而便捷的途径，育种家可根据需要从异源文库中选择目标区域渗入到栽培种的优良品系中，实现对栽培种目标性状的改良。目前，已有多个从事基因组测序、遗传资源和传统品种评估的研究联盟组织共同对番茄的多样性和适应性开展研究。

番茄参考基因组序列发布以来，很多新资源（基因组序列、数百万个 SNPs 等）、新研究工具（数据库、技术方法操作规程）和新研究方法（基因组编辑、作物模型和基因组选择）相继得到广泛应用，这必将使番茄育种工作变得更加高效。

对番茄生理过程、代谢途径、参与基因及候选基因的遗传变异性和突变的鉴定、翻译遗传学等方面研究的深入都将推动番茄育种的不断发展。未来番茄的种植条件如城镇园艺将是新的发展方向。

未来的番茄育种一定是建立在经验性方法与先进的生物技术相结合的基础上的，多学科结合也会变得愈来愈重要：①研发有效评价环境对作物影响作用的方法；②增强对性状的生物化学和分子基础的认识；③更好地理解基因型与环境互作，提高新品种对环境的适应性。

此外，还有一些复杂问题需要进一步研究，如几种胁迫如何互作、如何应对新的病虫害、根系与砧木如何互作、肥料如何科学减施等。建模将有助于将这几个方面综合起来设计出新的理想型番茄以适应多变的或优化的环境条件。

参考文献 REFERENCES

Aarts J, Hontelez J G J, Fischer P, et al., 1991. Acid phosphatase-11, a tightly linked molecular marker for root-knot nematode resistance in tomato—from protein to gene, using PCR and degenerate primers containing deoxyinosine. Plant Mol Biol, 16: 647-661.

Abraitiene A, Girgzdiene R, 2013. Impact of the short-term mild and severe ozone treatments on the potato spindle tuber viroid-infected tomato (*Lycopersicon esculentum* Mill.). Zemdirbyste-Agriculture, 100: 277-282.

Achuo E A, Prinsen E, Hofte M, 2006. Influence of drought, salt stress and abscisic acid on the resistance of tomato to Botrytis cinerea and *Oidium neolycopersici*. Plant Pathol, 55: 178-186.

Adams S R, Cockshull K E, Cave C R J, 2001. Effect of temperature on the growth and development of tomato fruits. Ann Bot, 88: 869-877.

Adams P, Ho L C, 1993. Effects of environment on the uptake and distribution of calcium in tomato and on the incidence of blossom-end rot. Plant Soil, 154: 127-132.

Adli M, 2018. The CRISPR tool kit for genome editing and beyond. Nat Commun, 9 (1): 1911.

Agrama H A, Scott J W, 2006. Quantitative trait loci for tomato yellow leaf curl virus and *tomato mottle virus* resistance in tomato. J Am Soc Hort Sci, 131: 267-272.

Ahmad A, Zhang Y, Cao X F, 2010. Decoding the epigenetic language of plant development. Mol Plant, 3: 719-728.

Al-Abdallat A, Al-Debei H, Ayad J, et al., 2014. Overexpression of *SlSHN1* gene improves drought tolerance by increasing cuticular wax accumulation in tomato. Int J Mol Sci, 15: 19499-19515.

Albacete A, Cantero-Navarro E, Großkinsky D K, et al., 2015. Ectopic overexpression of the cell wall invertase gene CIN1 leads to dehydration avoidance in tomato. J Exp Bot, 66: 863-878.

Albacete A, Martínez-Andújar C, Ghanem M E, et al., 2009. Rootstock-mediated changes in xylem ionic and hormonal status are correlated with delayed leaf senescence, and increased leaf area and crop productivity in salinized tomato. Plant Cell Environ, 32: 928-938.

参 考 文 献

Albert E, Duboscq R, Latreille M, et al., 2018. Allele-specific expression and genetic determinants of transcriptomic variations in response to mild water deficit in tomato. Plant J, 96 (3): 635-650.

Albert E, Gricourt J, Bertin N, et al., 2016. Genotype by watering regime interaction in cultivated tomato: lessons from linkage mapping and gene expression. Theor Appl Genet, 129: 395-418.

Albert E, Segura V, Gricourt J, et al., 2016. Association mapping reveals the genetic architecture of tomato response to water deficit: focus on major fruit quality traits. J Exp Bot, 67: 6413-6430.

Albrecht E, Escobar M, Chetelat R T, 2010. Genetic diversity and population structure in the tomatolike nightshades *Solanum lycopersicoides* and *S. sitiens*. Ann Bot, 105: 535-554.

Alian A, Altman A, Heuer B, 2000. Genotypic difference in salinity and water stress tolerance of fresh market tomato cultivars. Plant Sci, 152: 59-65.

Almeida J, Quadrana L, Asís R, et al., 2011. Genetic dissection of vitamin E biosynthesis in tomato. J Exp Bot, 62 (11): 3781-3798.

Alpert K B, Tanksley S D, 1996. High-resolution mapping and isolation of a yeast artificial chromosome contig containing *fw2.2*: a major fruit weight quantitative trait locus in tomato. Proc Natl Acad Sci USA, 93: 15503-15507.

Alseekh S, Fernie A R, 2018. Metabolomics 20 years on: what have we learned and what hurdles remain? Plant J, 94: 933-942.

Alseekh S, Ofner I, Pleban T, et al., 2013. Resolution by recombination: breaking up *Solanum pennellii* introgressions. Trends Plant Sci, 18: 536-538.

Alseekh S, Tong H, Scossa F, et al., 2017. Canalization of tomato fruit metabolism. Plant Cell, 29 (11): 2753-2765.

Ambros V, 2004. The functions of animal microRNAs. Nature, 431: 350-355.

Andolfo G, Jupe F, Witek K, et al., 2014. Defining the full tomato NB-LRR resistance gene repertoire using genomic and cDNA RenSeq. BMC Plant Biol, 14.

Anfoka G, Moshe A, Fridman L, et al., 2016. Tomato yellow leaf curl virus infection mitigates the heat stress response of plants grown at high temperatures. Sci Rep, 6: 19715.

Apse M P, Aharon G S, Snedden W A, et al., 1999. Salt tolerance conferred by overexpression of a vacuolar Na^+/H^+ antiport in Arabidopsis. Science, 285: 1256-1258.

Arafa R A, Rakha M T, Soliman N E K, et al., 2017. Rapid identification of candidate genes for resistance to tomato late blight disease using next-generation sequencing technologies. PLoS ONE, 12: e0189951.

Arms E M, Lounsbery J K, Bloom A J, et al., 2016. Complex relationships among water use efficiency-related traits, yield, and maturity in tomato lines subjected to deficit irrigation in the field. Crop Sci, 56: 1698.

Ashrafi H, Kinkade M P, Merk H L, et al., 2012. Identification of novel quantitative trait loci for increased lycopene content and other fruit quality traits in a tomato recombinant inbred line population. Mol Breed, 30: 549-567.

Ashrafi-Dehkordi E, Alemzadeh A, Tanaka N, et al., 2018. Meta-analysis of transcriptomic responses to biotic and abiotic stress in tomato. PeerJ, 6: e4631.

Asins M J, Albacete A, Martinez-Andujar C, et al., 2017. Genetic analysis of rootstock-mediated nitrogen (N) uptake and root-to-shoot signalling at contrasting N availabilities in tomato. Plant Sci, 263: 94-106.

Asins M J, Raga V, Roca D, et al., 2015. Genetic dissection of tomato rootstock effects on scion traits under moderate salinity. Theor Appl Genet, 128: 667-679.

Asins M J, Villalta I, Aly M M, et al., 2013. Two closely linked tomato HKT coding genes are positional candidates for the major tomato QTL involved in Na^+/K^+ homeostasis. Plant Cell Environ, 36: 1171-1191.

Atanassova B, 1999. Functional male sterility (*ps2*) in tomato (*Lycopersicon esculentum* Mill.) and its application in breeding and seed production. Euphytica, 107: 1, 13-21.

Auerswald H, Schwarz D, Kornelson C, et al., 1999. Sensory analysis, sugar and acid content of tomato at different EC values of the nutrient solution. Sci Hort (Amsterdam), 82: 227-242.

Bai Y, Lindhout P, 2007. Domestication and breeding of tomatoes: what have we gained and what can we gain in the future? Ann Bot, 100 (5): 1085-1094.

Bai Y L, Huang C C, Van Der Hulst R, et al., 2003. QTLs for tomato powdery mildew resistance (*Oidium lycopersici*) in *Lycopersicon parviflorum* G1.1601 colocalize with two qualitative powdery mildew resistance genes. Mol Plant-Microbe Interact, 16: 169-176.

Bai Y L, Kissoudis C, Yan Z, et al., 2018. Plant behaviour under combined stress: tomato responses to combined salinity and pathogen stress. Plant J, 93: 781-793.

Bai Y L, Pavan S, Zheng Z, et al., 2008. Naturally occurring broad-spectrum powdery mildew resistance in a central American tomato accession is caused by loss of Mlo function. Mol PlantMicrobe Interact, 21: 30-39.

Baldazzi V, Bertin N, Jong H, et al., 2012. Towards multiscale plant models: integrating cellular networks. Trends Plant Sci, 17: 728-736.

Baldazzi V, Génard M, Bertin N, 2017. Cell division, endoreduplication and expansion processes: setting the cell and organ control into an integrated model of tomato fruit development. Acta Hort, 1182.

参 考 文 献

Baldet P, Stevens R, Causse M, et al., 2007. Candidate genes and quantitative trait loci affecting fruit ascorbicacid content in three tomato populations. Plant Physiol, 143: 1943-1953.

Baldwin E, Scott J, Shewmaker C, et al., 2000. Flavor trivia and tomato aroma: biochemistry and possible mechanisms for control of important aroma components. Hort Science, 35: 1013-1022.

Baldwin E A, Nisperos-Carriedo M O, Baker R, et al., 1991. Quantitative analysis of flavor parameters in six Florida tomato cultivars (*Lycopersicon esculentum* Mill.). J Agri Food Chem, 39: 1135-1140.

Baldwin E A, Scott J W, Einstein M A, et al., 1998. Relationship between sensory and instrumental analysis for tomato flavor. J Am Soc Hort Sci, 123: 906-915.

Ballester A R, Bovy A G, Viquez-Zamora M, et al., 2016. Identification of loci affecting accumulation of secondary metabolites in tomato fruit of a *Solanum lycopersicum* × *Solanum chmielewskii* introgression line population. Front Plant Sci, 7: 1428.

Bandillo N, Raghavan C, Muyco P, et al., 2013. Multi-parent advanced generation intercross (MAGIC) populations in rice: progress and potential for genetics research and breeding. Rice, 6: 11.

Bauchet G, Grenier S, Samson N, et al., 2017. Use of modern tomato breeding germplasm for deciphering the genetic control of agronomical traits by Genome Wide Association study. Theor Appl Genet, 130: 875-889.

Bauchet G, Grenier S, Samson N, et al., 2017. Identification of major loci and genomic regions controlling acid and volatile content in tomato fruit: implications for flavor improvement. New Phytol, 215: 624-641.

Baxter C J, Liu J L, Fernie A R, et al., 2007. Determination of metabolic fluxes in a nonsteady-state system. Phytochemistry, 68: 2313-2319.

Beauvoit B, Belouah I, Bertin N, et al., 2018. Putting primary metabolism into perspective to obtain better fruits. Ann Bot, 122 (1): 1-21.

Beauvoit B P, Colombié S, Monier A, et al., 2014. Model-assisted analysis of sugar metabolism throughout tomato fruit development reveals enzyme and carrier properties in relation to vacuole expansion. Plant cell, 26 (8): 3224-3242.

Bernacchi D, Beck-Bunn T, Emmatty D, et al., 1998. Advanced backcross QTL analysis in tomato. II. Evaluation of nearisogenic lines carrying single-donor introgressions for desirable wild QTL-alleles derived from *Lycopersicon hirsutum* and *L. pimpinellifolium*. Theor Appl Genet, 97 (1/2): 170-180; erratum 97 (7): 1191-1196.

Bernacchi D, Beck-Bunn T, Eshed Y, et al., 1998. Advanced backcross QTL analysis in tomato. I. Identification of QTLs for traits of agronomic importance from *Lycopersicon hirsutum*. Theor Appl Genet, 97: 381-397.

Berr A, Shafiq S, Shen W H, 2011. Histone modifications in transcriptional activation during plant development. Biochim Biophys Acta Gene Regul Mech 1809: 567-576.

Bertin N, Borel C, Brunel B, et al., 2003. Do genetic make-up and growth manipulation affect tomato fruit size by cell number, or cell size and DNA endoreduplication? Ann Bot, 92 (3): 415-424.

Bertin N, Guichard S, Leonardi C, et al., 2000. Seasonal evolution of the quality of fresh glasshouse tomatoes under mediterranean conditions, as affected by air vapour pressure deficit and plant fruit load. Ann Bot, 85: 741-750.

Bertin N, Gautier H, Roche C, 2002. Number of cells in tomato fruit depending on fruit position and source-sink balance during plant development. Plant Growth Regul, 36 (2): 105-112.

Bertin N, Martre P, Génard M, et al., 2010. Why and how can process-based simulation models link genotype to phenotype for complex traits? Case-study of fruit and grain quality traits. J Exp Bot, 61: 955-967.

Bhatia P, Ashwath N, Senaratna T, et al., 2004. Tissue culture studies of tomato (*Lycopersicon esculentum*). Plant Cell Tiss Org Cult, 78 (1): 1-21.

Bhatt R M, Srinivasa Rao N K, 1987. Seed germination and seedling growth responses of tomato cultivars to imposed water stress. JHort Sci, 62: 221-225.

Birchler J A, Yao H, Chudalayandi S, et al., 2010. Heterosis. Plant Cell, 22: 2105-2112.

Blanca J, Cañizares J, Cordero L, et al., 2012. Variation revealed by SNP genotyping and morphology provides insight into the origin of the tomato. PLoS ONE, 7: e48198.

Blanca J, Montero-Pau J, Sauvage C, et al., 2015. Genomic variation in tomato, from wild ancestors to contemporary breeding accessions. BMC Genom, 16: 257.

Bloom A J, Zwieniecki M A, Passioura J B, et al., 2004. Water relations under root chilling in a sensitive and tolerant tomato species. Plant, Cell Environ, 27: 971-979.

Boison S A, Utsunomiya A T H, Santos D J A, et al., 2017. Accuracy of genomic predictions in Gyr (Bos indicus) dairy cattle. J Dairy Sci, 100: 5479-5490.

Bolger A, Scossa F, Bolger M E, et al., 2014. The genome of the stress-tolerant wild tomato species *Solanum pennellii*. Nat Genet, 46: 1034-1038.

Boureau L, How-Kit A, Teyssier E, et al., 2016. A CURLY LEAF homologue controls both vegetative and reproductive development of tomato plants. Plant Mol Biol, 90: 485-501.

Bovy A, De Vos R, Kemper M, et al., 2002. High-flavonol tomatoes resulting from the heterologous expression of the maize transcription factor genes LC and C1. Plant Cell, 14: 2509-2526.

参 考 文 献

Bovy A, Schijlen E, Hall R D, 2007. Metabolic engineering of flavonoids in tomato (*Solanum lycopersicum*): the potential for metabolomics. Metabolomics, 3: 399-412.

Boote K J, Rybak M R, Scholberg J M, et al., 2012. Improving the CROPGRO-Tomato model for predicting growth and yield response to temperature. HortScience, 47: 1038-1049.

Brachi B, Morris G P, Borevitz J O, 2011, Genome-wide association studies in plants: the missing heritability is in the field. Genome Biol 12: 232 Bramley PM (2000) Is lycopene beneficial to human health? Phytochemistry, 54: 233-236.

Brandwagt B F, Mesbah L A, Takken F L W, et al., 2000. A longevity assurance gene homolog of tomato mediates resistance to *Alternaria alternata* f. sp *lycopersici* toxins and fumonisin B (1). Proceedings of the national academy of sciences of the United States of America, 97: 4961-4966.

Breiman L, 2001. Random forests. Mach Learn, 45: 5-32.

Brommonschenkel S H, Frary A, Tanksley S D, 2000. The broad-spectrum tospovirus resistance gene *Sw-5* of tomato is a homolog of the root-knot nematode resistance gene *Mi*. Mol Plant-Microbe Interact, 13: 1130-1138.

Brooks C, Nekrasov V, Lippman Z B, et al., 2014. Efficient gene editing in tomato in the first generation using the clustered regularly interspaced short palindromic repeats/CRISPRassociated9 system. Plant Physiol, 166 (3): 1292-1297.

Brouwer D J, Jones E S, St Clair D A, 2004. QTL analysis of quantitative resistance to *Phytophthora infestans* (late blight) in tomato and comparisons with potato. Genome, 47: 475-492.

Brouwer D J, St Clair D A, 2004. Fine mapping of three quantitative trait loci for late blight resistance in tomato using near isogenic lines (NILs) and sub-NILs. Theor Appl Genet, 108: 628-638.

Browning B L, Browning S R, 2016. Genotype imputation with millions of reference samples. Amer J Hum Genetm, 98: 116-126.

Bruhn C M, Feldman N, Garlitz C, et al., 1991. Consumer perceptions of quality: apricots, cantaloupes, peaches, pears, strawberries, and tomatoes. J Food Qual, 14: 187-195.

Brummell D A, Harpster M H, Civello P M, et al., 1999. Modification of expansin protein abundance in tomato fruit alters softening and cell wall polymer metabolism during ripening. Plant Cell, 11 (11): 2203-2216.

Bucheli P, Voirol E, De La Torre R, et al., 1999. Definition of nonvolatile markers for flavor of tomato (*Lycopersicon esculentum* Mill.) as tools in selection and breeding. J Agri Food Chem, 47: 659-664.

Budiman M A, Chang S B, Lee S, et al., 2004. Localization of *jointless-2* gene in the centromeric region of tomato chromosome 12 based on high resolution genetic and physical mapping. Theor Appl Genet, 108: 190-196.

Bush D S, 1995. Calcium regulation in plant cells and its role in signaling. Annu Rev Plant Physiol, 46: 95-122.

Bussières P, 1994. Water import rate in tomato fruit: a resistance model. Ann Bot, 73: 75-82.

Butler L, 1952. The linkage map of the tomato. J Hered, 43: 25-36.

Cagas C C, Lee O N, Nemoto K, et al., 2008. Quantitative trait loci controlling flowering time and related traits in a *Solanum lycopersicum* × *S. pimpinellifolium* cross. Sci Hort (Amsterdam), 116: 144-151.

Calin G A, Croce C M, 2006. MicroRNA signatures in human cancers. Nat Rev Cancer, 6: 857-866.

Callaway E, 2018. CRISPR plants now subject to tough GM laws in European Union. Nature, 560: 16.

Calus M P L, Meuwissen T H E, De Roos APW, et al., 2008. Accuracy of genomic selection using different methods to define haplotypes. Genetics, 178: 553-561.

Canady M A, Meglic V, Chetelat R T, 2005. A library of Solanum lycopersicoides introgression lines in cultivated tomato. Genome, 48: 685-697.

Cantero-Navarro E, Romero-Aranda R, Fernández-Muñoz R, et al., 2016. Improving agronomic water use efficiency in tomato by rootstock-mediated hormonal regulation of leaf biomass. Plant Sci, 251: 90-100.

Cárdenas P D, Sonawane P D, Pollier J, et al., 2016. *GAME9* regulates the biosynthesis of steroidal alkaloids and upstream isoprenoids in the plant mevalonate pathway. Nat Commun, 7: 10654.

Carelli B P, Gerald L T S, Grazziotin F G, et al., 2006. Genetic diversity among Brazilian cultivars and landraces of tomato *Lycopersicon esculentum* Mill. revealed by RAPD markers. Genet Resour Crop Evol, 53: 395-400.

Carmeille A, Caranta C, Dintinger J, et al., 2006. Identification of QTLs for *Ralstonia solanacearum* race 3-phylotype II resistance in tomato. Theor Appl Genet, 113: 110-121.

Carmel-Goren L, Liu Y S, Lifschitz E, et al., 2003. The SELF-PRUNING gene family in tomato. Plant Mol Biol, 52: 1215-1222.

Caro M, Cruz V, Cuartero J, et al., 1991. Salinity tolerance of normal-fruited and cherry tomato cultivars. Plant Soil, 136: 249-255.

Caromel B, Hamers C, Touhami N, et al., 2015. Screening tomato germplasm for resistance to late blight. In: INNOHORT, innovation in integrated & organic horticulture. ISHS International Symposium, Avignon, France, 15-16.

Carrari F, Baxter C, Usadel B, et al., 2006. Integrated analysis of metabolite and transcript levels reveals the metabolic shifts that underlie tomato fruit development and highlight regulatory aspects of metabolic network behavior. Plant Physiol, 142: 1380-1396.

Casteel C L, Walling L L, Paine T D, 2007. Effect of *Mi-1.2* gene in natal host plants on behavior and biology of the tomato psyllid *Bactericerca cockerelli* (Sulc) (Hemiptera: Psyllidae). J Entomol Sci, 42: 155-162.

Catanzariti A M, Lim G T T, Jones D A, 2015. The tomato *I-3* gene: a novel gene for resistance to Fusarium wilt disease. New Phytol, 207: 106-118.

Catchen J M, Boone J Q, Davey J W, et al., 2011. Genome-wide genetic marker discovery and genotyping using next-generation sequencing. Nat Rev Genet, 12: 499-510.

Causse M, Buret M, Robini K, et al., 2003. Inheritance of nutritional and sensory quality traits in fresh market tomato and relation to consumer preferences. J Food Sci, 68: 2342-2350.

Causse M, Friguet C, Coiret C, et al., 2010. Consumer preferences for fresh tomato at the European scale: a common segmentation on taste and firmness. J Food Sci, 75 (9): 531-541.

Causse M, Chaïb J, Lecomte L, et al., 2007. Both additivity and epistasis control the genetic variation for fruit quality traits in tomato. Theor Appl Genet, 115: 429-442.

Causse M, Duffe P, Gomez M C, et al., 2004. A genetic map of candidate genes and QTLs involved in tomato fruit size and composition. J Exp Bot, 55: 1671-1685.

Causse M, Damidaux R, Rousselle P, 2007. Traditional and enhanced breeding for fruit quality traits in tomato. Enfield Science Publishers.

Causse M, Saliba-Colombani V, Lecomte L, et al., 2002. QTL analysis of fruit quality in fresh market tomato: a few chromosome regions control the variation of sensory and instrumental traits. J Exp Bot, 53: 2089-2098.

Causse M, Desplat N, Pascual L, et al., 2013. Whole genome resequencing in tomato reveals variation associated with introgression and breeding events. BMC Genomics, 14: 791.

Chakrabarti M, Zhang N, Sauvage C, et al., 2013. A cytochrome P450 regulates a domestication trait in cultivated tomato. Proc Natl Acad Sci USA, 110: 17125-17130.

Chen F Q, Foolad M R, Hyman J, et al., 1999. Mapping of QTLs for lycopene and other fruit traits in a *Lycopersicon esculentum* × *L. pimpinellifolium* cross and comparison of QTLs across tomato species. Mol Breed, 5: 283-299.

Chen J, Kang S, Du T, et al., 2013. Quantitative response of greenhouse tomato yield and quality to water deficit at different growth stages. Agri Water Manag, 129: 152-162.

Chen X, 2005. MicroRNA biogenesis and function in plants. FEBS Lett, 579: 5923.

Chen X, 2009. Small RNAs and their roles in plant development. Annu Rev Cell Dev Biol, 25: 21-44.

Chetelat R T, De Verna J W, Bennett A B, 1995. Introgression into tomato (*Lycopersicon esculentum*) of the *L. chmielewskii* sucrose accumulator gene (*sucr*) controlling fruit sugar composition. Theor Appl Genet, 91: 327-333.

Cho S K, Ben C S, Shah P, et al., 2014. COP1 E3 ligase protects HYL1 to retain microRNA biogenesis. Nat Commun, 5: 5867.

Chunwongse J, Chunwongse C, Black L, et al., 2002. Molecular mapping of the *Ph-3* gene for late blight resistance in tomato. J Hort Sci Biotechnol, 77: 281-286.

Clark A G, 2004. The role of haplotypes in candidate gene studies. Genet Epidemiol, 27: 321-333.

Clough S J, Bent A F, 1998. Floral dip: a simplified method for Agrobacterium-mediated transformation of Arabidopsis thaliana. Plant J, 16 (6): 735-743.

Coaker G L, Francis D M, 2004. Mapping, genetic effects, and epistatic interaction of two bacterial canker resistance QTLs from Lycopersicon hirsutum. Theor Appl Genet, 108: 1047-1055.

Colliver S, Bovy A, Collins G, et al., 2002. Improving the nutritional content of tomatoes through reprogramming their flavonoid biosynthetic pathway. Phytochem Rev, 1: 113-123.

Colombié S, Beauvoit B, Nazaret C, et al., 2017. Respiration climacteric in tomato fruits elucidated by constraint-based modelling. New Phytol, 213: 1726-1739.

Colombié S, Nazaret C, Bénard C, et al., 2015. Modelling central metabolic fluxes by constraint-based optimization reveals metabolic reprogramming of developing *Solanum lycopersicum* (tomato) fruit. Plant J, 81: 24-39.

Comai L, Henikoff S, 2006. TILLING: practical single-nucleotide mutation discovery. Plant J, 45: 684-694.

Coneva V, Frank M H, Balaguer M A L, et al., 2017. Genetic architecture and molecular networks underlying leaf thickness in desert-adapted tomato *Solanum pennellii*. Plant Physiol, 175 (1): 376-391.

Constantinescu D, Memmah M M, Vercambre G, et al., 2016. Model-assisted estimation of the genetic variability in physiological parameters related to tomato fruit growth under contrasted water conditions. Front Plant Sci, 7: 1841.

Costa J M, Ortuño M F, Chaves M M, 2007. Deficit irrigation as a strategy to save water: physiology and potential application to horticulture. J Integr Plant Biol, 49: 1421-1434.

Cournède P H, et al., 2013. Development and evaluation of plant growth models:

methodology and implementation in the pygmalion platform. Math Mod Nat Phen, 8 (4): 112-130.

Cowger C, Brown J K M, 2019. Durability of quantitative resistance in crops: greater than we know? Annu Rev Phytopathol, 57: 253-277.

Crain J, Mondal S, Rutkoski J, et al. , 2018. Combining high-Throughput phenotyping and genomic information to increase prediction and selection accuracy in wheat breeding. Plant Genome, 11: 20.

Crossa J, Pérez-Rodríguez P, Cuevas J, et al. , 2017. Genomic selection in plant breeding: methods, models, and perspectives. Trends Plant Sci.

Cui J, Jiang N, Zhou X, et al. , 2018. Tomato MYB49 enhances resistance to *Phytophthora infestans* and tolerance to water deficit and salt stress. Planta, 248: 1487-1503.

Cui J, You C, Chen X, 2017. The evolution of microRNAs in plants. Curr Opin Plant Biol, 35: 61-67.

Cui J, Zhou B, Ross S A, et al. , 2017. Nutrition, microRNAs, and human health. Adv Nutr, 8: 105-112.

Cuyabano B C D, Su G, Lund M S, 2015. Selection of haplotype variables from a high-density marker map for genomic prediction. Genet Sel Evol, 47: 61.

Cuyabano B C D, Su G, Rosa G J M, et al. , 2015. Bootstrap study of genomeenabled prediction reliabilities using haplotype blocks across Nordic Red cattle breeds. J Dairy Sci, 98: 7351-7363.

Dal Cin V, Kevany B, Fei Z, et al. , 2009. Identification of *Solanum habrochaites* loci that quantitatively influence tomato fruit ripening-associated ethylene emissions. Theor Appl Genet, 119: 1183-1192.

Danecek P, Huang J, Min J L, et al. , 2015. Improved imputation of low-frequency and rare variants using the UK10K haplotype reference panel. Nat Commun, 6: 8111.

Danilo B, Perrot L, Botton E, et al. , 2018. The DFR locus: a smart landing pad for targeted transgene insertion in tomato. PLoS ONE, 13 (12): e0208395.

Danilo B, Perrot L, Mara K, et al. , 2019. Efficient and transgene-free gene targeting using Agrobacterium-mediated delivery of the CRISPR/Cas9 system in tomato. Plant Cell Rep, 38 (4): 459-462.

Das S, Forer L, Schönherr S, et al. , 2016. Next-generation genotype imputation service and methods. Nat Genet, 48: 1284-1287.

Davies J N, Hobson G E, 1981. The constituents of tomato fruit—the influence of environment, nutrition, and genotype. Crit Rev Food SciNutr, 15: 205-280.

Davila O N H, Kruijer W, Gort G, et al. , 2017. Genome-wide association analysis reveals distinct genetic architectures for single and combined stress responses in Arabidopsis thaliana. New Phytol, 213: 838-851.

Davis J, Yu D Z, Evans W, et al., 2009. Mapping of loci from *Solanum lycopersicoides* conferring resistance or susceptibility to Botrytis cinerea in tomato. Theor Appl Genet, 119: 305-314.

De Freitas S T, Martinelli F, Feng B, et al., 2018. Transcriptome approach to understand the potential mechanisms inhibiting or triggering blossom-end rot development in tomato fruit in response to plant growth regulators. J Plant GrowthRegul, 37: 183-198.

De Groot C C, Marcelis L F M, van den Boogaard R, et al., 2004. Response of growth of tomato to phosphorus and nitrogen nutrition. Acta Hort, 357-364.

De Jong C F, Takken F L W, Cai X H, et al., 2002. Attenuation of Cf-mediated defense responses at elevated temperatures correlates with a decrease in elicitor-binding sites. Mol PlantMicrobe Interact, 15: 1040-1049.

De Los Campos G, Hickey J M, Pong-Wong R, et al., 2013. Whole-genome regression and prediction methods applied to plant and animal breeding. Genetics, 193: 327-345.

De Swaef T, Mellisho C D, Baert A, et al., 2014. Model-assisted evaluation of crop load effects on stem diameter variations and fruit growth in peach. Trees, 28: 1607-1622.

Delhaize E, Gruber B D, Ryan P R, 2007. The roles of organic anion permeases in aluminium resistance and mineral nutrition. FEBS Lett, 581: 2255-2262.

De Vicente M C, Tanksley S D, 1993. QTL analysis of transgressive segregation in an interspecific tomato cross. Genetics, 134: 585-596.

Dileo M V, Pye M F, Roubtsova T V, et al., 2010. Abscisic acid in salt stress predisposition to Phytophthora root and crown rot in tomato and chrysanthemum. Phytopathology, 100: 871-879.

Diouf I A, Derivot L, Bitton F, et al., 2018. Water deficit and salinity stress reveal many specific QTL for plant growth and fruit quality traits in tomato. Front Plant Sci, 9: 279.

Dixon M S, Hatzixanthis K, Jones D A, et al., 1998. The tomato Cf-5 disease resistance gene and six homologs show pronounced allelic variation in leucine-rich repeat copy number. Plant Cell, 10: 1915-1925.

Dixon M S, Jones D A, Keddie J S, et al., 1996. The tomato *Cf-2* disease resistance locus comprises two functional genes encoding leucine-rich repeat proteins. Cell, 84: 451-459.

Do P T, Prudent M, Sulpice R, et al., 2010. The influence of fruit load on the tomato pericarp metabolome in a *Solanum chmielewskii* introgression line population. Plant Physiol, 154: 1128-1142.

Doganlar S, Dodson J, Gabor B, et al., 1998. Molecular mapping of the *py-1* gene for resistance to corky root rot (*Pyrenochaeta lycopersici*) in tomato. Theor Appl Genet, 97: 784-788.

参 考 文 献

Doganlar S, Frary A, Ku H M, et al., 2003. Mapping quantitative trait loci in inbred backcross lines of *Lycopersicon pimpinellifolium* (LA1589). Genome, 45: 1189-1202.

Donald C M, 1968. The breeding of cropidéotypes. Euphytica, 17: 385-403.

Dong Q L, Liu D D, An X H, et al., 2011. *MdVHP1* encodes an apple vacuolar H^+-PPase and enhances stress tolerance in transgenic apple callus and tomato. JPlant Physiol, 168 (17): 2124-2133.

Dong Z, Men Y, Li Z, et al., 2019. Chlorophyll fluorescence imaging as a tool for analyzing the effects of chilling injury on tomato seedlings. SciHort (Amsterdam), 246: 490-497.

Dorais M, Papadopoulos A P, Gosselin A, 2001. Greenhouse tomato fruit quality. Hortic Rev, 26: 239-319.

Dreissig S, Schiml S, Schindele P, et al., 2017. Live-cell CRISPR imaging in plants reveals dynamic telomere movements. Plant J, 91 (4): 565-573.

Driedonks N, Wolters-Arts M, Huber H, et al., 2018. Exploring the natural variation for reproductive thermotolerance in wild tomato species. Euphytica, 214: 67.

Du Y D, Niu W Q, Gu X B, et al., 2018. Water -and nitrogen-saving potentials in tomato production: a meta-analysis. Agri Water Manag, 210: 296-303.

Duangjit J, Causse M, Sauvage C, 2016. Efficiency of genomic selection for tomato fruit quality. Mol Breed, 36 (36): 29.

Edwards S M, Buntjer J B, Jackson R, et al., 2019. The effects of training population design on genomic prediction accuracy in wheat. Theor Appl Genet, 443267.

El-hady E, Haiba A, El-hamid N R A, et al., 2010. Phylogenetic diversity and relationships of some tomato varieties by electrophoretic protein and RAPD analysis. J Amer Sci, 6: 434-441.

Elvanidi A, Katsoulas N, Augoustaki D, et al., 2018. Crop reflectance measurements for nitrogen deficiency detection in a soilless tomato crop. Biosyst Eng, 176: 1-11.

Endelman J B, 2011. Ridge regression and other kernels for genomic selection with R package rrBLUP. Plant Genome J, 4: 250.

Ercolano M R, Sanseverino W, Carli P, et al., 2012. Genetic and genomic approaches for R-gene mediated disease resistance in tomato: retrospects and prospects. Plant Cell Rep, 31: 973-985.

Eriksson E M, Bovy A, Manning K, et al., 2004. Effect of the colorless non-ripening mutation on cell wall biochemistry and gene expression during tomato fruit development and ripening 1 [w]. Plant Physiol, 136: 4184-4197.

Ernst K, Kumar A, Kriseleit D, et al., 2002. The broad-spectrum potato cyst nematode resistance gene (*Hero*) from tomato is the only member of a large gene family of NBS-

LRR genes with an unusual amino acid repeat in the LRR region. Plant J, 31: 127-136.

Eshed Y, Gera G, Zamir D, 1996. A genome-wide search for wild-species alleles that increase horticultural yield of processing tomato. Theor Appl Genet, 93: 877-886.

Eshed Y, Zamir D, 1995. An introgression line population of *Lycopersicon pennellii* in the cultivatedtomato enables the identification and fine mapping of yield-associated QTL. Genetics, 141: 1147-1162.

Estañ M T, Villalta I, Bolarín M C, et al., 2009. Identification of fruit yield loci controlling the salt tolerance conferred by solanum rootstocks. Theor Appl Genet, 118: 305-312.

Evangelou E, Ioannidis J P A, 2013. Meta-analysis methods for genome-wide association studies and beyond. Nat Rev Genet, 14: 379-389.

Fan Z Q, Ba L J, Shan W, et al., 2018. A banana *R2R3-MYB* transcription factor *MaMYB3* is involved in fruit ripening through modulation of starch degradation by repressing starch degradation-related genes and *MabHLH6*. Plant J, 96 (6): 1191-1205.

Fang X, Cui Y, Li Y, et al., 2015. Transcription and processing of primary microRNAs are coupled by Elongator complex in Arabidopsis. Nat Plants, 1: 15075.

Fanwoua J, De Visser P H B, Heuvelink E, et al., 2013. A dynamic model of tomato fruit growth integrating cell division, cell growth and endoreduplication. Funct Plant Biol, 40 (11): 1098-1114.

Farashi S, Kryza T, Clements J, et al., 2019. Post-GWAS in prostate cancer: from genetic association to biological contribution. Nat Rev Cancer, 19: 46-59.

Fereres E, Soriano M A, 2006. Deficit irrigation for reducing agricultural water use. J Exp Bot, 58: 147-159.

Fernandes S B, Dias K O G, Ferreira D F, et al., 2018. Efficiency of multi-trait, indirect, and traitassisted genomic selection for improvement of biomass sorghum. Theor Appl Genet, 131: 747-755.

Fernandez A I, Viron N, Alhagdow M, et al., 2009. Flexible tools for gene expression and silencing in tomato. Plant Physiol, 151 (4): 1729-1740.

Fernie A R, Aharoni A, Willmitzer L, et al., 2011. Recommendations for reporting metabolite data. Plant Cell, 23: 2477-2482.

Fernie A R, Schauer N, 2009. Metabolomics-assisted breeding: a viable option for crop improvement? Trends Genet, 25: 39-48.

Finkers R, Bai Y L, Van den Berg P, et al., 2008. Quantitative resistance to Botrytis cinerea from *Solanum neorickii*. Euphytica, 159: 83-92.

Finkers R, Van Den Berg P, Van Berloo R, et al., 2007. Three QTLs for Botrytis cinerea resistance in tomato. Theor Appl Genet, 114: 585-593.

Finkers R, Van Heusden A W, Meijer-Dekens F, et al., 2007. The construction of a

Solanum habrochaites LYC4 introgression line population and the identification of QTLs for resistance to Botrytis cinerea. Theor Appl Genet, 114: 1071-1080.

Fishman S, Génard M, 1998. A biophysical model of fruit growth: simulation of seasonal and diurnal dynamics of mass. Plant Cell Environ 21: 739-752.

Foolad M R, Merk H L, Ashrafi H, 2008. Genetics, genomics and breeding of late blight and early blight resistance in tomato. Crit Rev Plant Sci, 27: 75-107.

Foolad M R, Panthee D R, 2012. Marker-assisted selection in tomato breeding. Crit Rev Plant Sci, 31: 93-123.

Foolad M R, Sullenberger M T, Ohlson E W, et al., 2014. Response of accessions within tomato wild species, *Solanum pimpinellifolium* to late blight. Plant Breed, 133: 401-411.

Foolad M R, Zhang L P, Khan A A, et al., 2002. Identification of QTLs for early blight (*Alternaria solani*) resistance in tomato using backcross populations of a *Lycopersicon esculentum* × *L. hirsutum* cross. Theor Appl Genet, 104: 945-958.

Fragkostefanakis S, Mesihovic A, Simm S, et al., 2016. *HsfA2* controls the activity of developmentally and stress regulated heat stress protection mechanisms in tomato male reproductive tissues. Plant Physiol, 170: 2461-2477.

Fragkostefanakis S, Röth S, Schleiff E, et al., 2015. Prospects of engineering thermotolerance in crops through modulation of heat stress transcription factor and heat shock protein networks. Plant, Cell Environ, 38: 1881-1895.

Frary A, Doganlar S, Daunay M C, et al., 2003. QTL analysis of morphological traits in eggplant and implications for conservation of gene function during evolution of solanaceous species. Theor Appl Genet, 107: 359-370.

Frary A, Fulton T M, Zamir D, et al., 2004. Advanced backcross QTL analysis of a *Lycopersicon esculentum* × *L. pennellii* cross and identification of possible orthologs in the Solanaceae. Theor Appl Genet, 108: 485-496.

Frary A, Kele S D, Pinar H, et al., 2011. NaCl tolerance in *Lycopersicon pennellii* introgression lines: QTL related to physiological responses. Biol Plant, 55: 461-468.

Frary A, Nesbitt T C, Frary A, et al., 2000. *fw2.2*: a quantitative trait locus key to the evolution of tomato fruit size. Science, (80-) 289: 85-88.

Frary A, Göl D, Kele S D, et al., 2010. Salt tolerance in *Solanum pennellii*: antioxidant response and related QTL. BMC Plant Biol, 10: 58.

Fridman E, Carrari F, Liu Y S, et al., 2004. Zooming in on a quantitative trait for tomato yield using interspecific introgressions. Science, (80-) 305: 1786-1789.

Fridman E, Liu Y S, Carmel-Goren L, et al., 2002. Two tightly linked QTLs modify tomato sugar content via different physiological pathways. Mol Genet Genom, 266:

821-826.

Fridman E, Pleban T, Zamir D, 2000. A recombination hotspot delimits a wild-species quantitative trait locus for tomato sugar content to 484 bp within an invertase gene. Proc Natl Acad Sci USA, 97: 4718-4723.

Fridman E, Zamir D, 2003. Functional divergence of a syntenic invertase gene family in tomato, potato, and Arabidopsis. Plant Physiol, 131: 603-609.

Fry W E, Goodwin S B, 1997. Re-emergence of potato and tomato late blight in the United States. Plant Dis, 81: 1349-1357.

Fujita M, Fujita Y, Noutoshi Y, et al., 2006. Crosstalk between abiotic and biotic stress responses: a current view from the points of convergence in the stress signaling networks. Curr Opin Plant Biol, 9: 436-442.

Fulop D, Ranjan A, Ofner I, et al., 2016. A new advanced backcross tomato population enables high resolution leaf QTL mapping and gene identification. G3: Genes Genomes Genet, 6: 3169-3184.

Fulton T M, 2002. Identification, analysis, and utilization of conserved ortholog set markers for comparative genomics in higher plants. Plant Cell, 14: 1457-1467.

Fulton T M, Beck-Bunn T, Emmatty D, et al., 1997. QTL analysis of an advanced backcross of *Lycopersicon peruvianum* to the cultivated tomato and comparisons with QTLs found in other wild species. Theor Appl Genet, 95: 881-894.

Fulton T M, Grandillo S, Beck-Bunn T, et al., 2000. Advanced backcross QTL analysis of a *Lycopersicon esculentum* × *Lycopersicon parviflorum* cross. Theor Appl Genet, 100: 1025-1042.

Gaion L A, Muniz J C, Barreto R F, et al., 2019. Amplification of gibberellins response in tomato modulates calcium metabolism and blossom end rot occurrence. Sci Hortic (Amsterdam), 246: 498-505.

Gaj T, Gersbach C A, Barbas C F, 2013. ZFN, TALEN, and CRISPR/Cas-based methods for genome engineering. Trends Biotechnol, 31 (7): 397-405.

Gallusci P, Hodgman C, Teyssier E, et al., 2016. DNA methylation and chromatin regulation during fleshy fruit development and ripening. Front Plant Sci, 7: 807.

Gao C, Ju Z, Cao D, et al., 2015. MicroRNA profiling analysis throughout tomato fruit development and ripening reveals potential regulatory role of RIN on microRNAs accumulation. Plant Biotechnol J, 13: 370-382.

Gao L, Gonda I, Sun H, et al., 2019. The tomato pan-genome uncovers new genes and a rare allele regulating fruit flavor. Nat Genet, 51: 1044-1051.

Garcia V, Bres C, Just D, et al., 2016. Rapid identification of causal mutations in tomato EMS populations via mapping-by-sequencing. Nat Protoc, 11: 2401-2418.

Gauffier C, Lebaron C, Moretti A, et al., 2016. A TILLING approach to generate broad-spectrum resistance to potyviruses in tomato is hampered by *eIF4E* gene redundancy. Plant J, 85: 717-729.

Gautier H, Diakou-Verdin V, Bénard C, et al., 2008. How does tomato quality (sugar, acid, and nutritional quality) vary with ripening stage, temperature, and irradiance? J Agri Food Chem, 56: 1241-1250.

Génard M, Bertin N, Gautier H, et al., 2010. Virtual profiling: a new way to analyse phenotypes. Plant J, 62: 344-355.

Génard M, Lescourret F, 2004. Modelling fruit quality: ecophysiological, agronomical and ecological perspectives. Netherlands: Kluwer Academic Publisher.

Gerszberg A, Hnatuszko-Konka K, Kowalczyk T, et al., 2015. Tomato (*Solanum lycopersicum* L.) in the service of biotechnology. Plant Cell Tiss Org Cult, 120 (3): 881-902.

Geshnizjani N, Ghaderi-Far F, Willems L A J, et al., 2018. Characterization of and genetic variation for tomato seed thermo-inhibition and thermo-dormancy. BMC Plant Biol, 18: 229.

Gest N, Gautier H, Stevens R, 2013. Ascorbate as seen through plant evolution: the rise of a successful molecule? J Exp Bot, 64: 33-53.

Gianola D, van Kaam J B C H M, 2008. Reproducing Kernel Hilbert spaces regression methods for genomic assisted prediction of quantitative traits. Genetics, 178: 2289.

Giovannoni J, Nguyen C, Ampofo B, et al., 2017. The epigenome and transcriptional dynamics of fruit ripening. Annu Rev Plant Biol, 68: 61-84.

Giovannucci E, 1999. Tomatoes, tomato-based products, lycopene, and cancer: review of the epidemiologic literature. J Natl Cancer Inst, 91: 317-331.

Giroux R W, Filion W G, 1992. A comparison of the chilling-stress response in two differentially tolerant cultivars of tomato (*Lycopersicon esculentum*). Biochem Cell Biol, 70: 191-198.

Goff S A, Klee H J, 2006. Plant volatile compounds: sensory cues for health and nutritional value? Science, 311: 815-819.

Gonatopoulos-Pournatzis T, Cowling V H, 2015. Cap-binding complex (CBC). Biochem J, 458: 185.

Gonzalez-Cendales Y, Catanzariti A M, Baker B, et al., 2016. Identification of *I-7* expands the repertoire of genes for resistance to Fusarium wilt in tomato to three resistance gene classes. Mol Plant Pathol, 17: 448-463.

Goodwin S, McPherson J D, McCombie W R, 2016. Coming of age: ten years of next-generation sequencing technologies. Nat Rev Genet, 17: 333-351.

Grandillo S, Ku H M, Tanksley S D, 1996. Characterization of *fs8.1*, a major QTL influencing fruit shape in tomato. Mol Breed, 2: 251-260.

Grandillo S, Ku H M, Tanksley S D, 1999. Identifying the loci responsible for natural variation in fruit size and shape in tomato. Theor Appl Genet, 99: 978-987.

Grandillo S, Tanksley S D, 1996. Genetic analysis of *RFLPs*, *GATA* microsatellites and *RAPDs* in a cross between *L. esculentum* and *L. pimpinellifolium*. Theor Appl Genet, 92: 957-965.

Grandillo S, Tanksley S D, 1996. QTL analysis of horticultural traits differentiating the cultivated tomato from the closely related species *Lycopersicon pimpinellifolium*. Theor Appl Genet, 92: 935-951.

Grierson D, 2016. Identifying and silencing tomato ripening genes with antisense genes. Plant Biotechnol J, 14 (3): 835-838.

Grilli G, Trevizan Braz L, Gertrudes E, et al., 2007. QTL identification for tolerance to fruit set in tomato by *fAFLP* markers. Crop Breed Appl Biotechnol, 7: 234-241.

Grimson A, Srivastava M, Fahey B, et al., 2008. Early origins and evolution of microRNAs and Piwi-interacting RNAs in animals. Nature, 455: 1193-1197.

Guan Y, Stephens M, 2008. Practical issues in imputation-based association mapping. PLoS Genet.

Guichard S, Bertin N, Leonardi C, et al., 2001. Tomato fruit quality in relation to water and carbon fluxes. Agronomie, 21: 385-392.

Guichard S, Gary C, Leonardi C, et al., 2005. Analysis of growth and water relations of tomato fruits in relation to air vapor pressure deficit and plant fruit load. J Plant Growth Regul, 24: 201-213.

Gur A, Osorio S, Fridman E, et al., 2010. hi2-1, a QTL which improves harvest index, earliness and alters metabolite accumulation of processing tomatoes. Theor Appl Genet, 121: 1587-1599.

Gur A, Semel Y, Osorio S, et al., 2011. Yield quantitative trait loci from wild tomato are predominately expressed by the shoot. Theor Appl Genet, 122: 405-420.

Haanstra J P W, Wye C, Verbakel H, et al., 1999. An integrated high-density RFLP-AFLP map of tomato based on two *Lycopersicon esculentum* × *L. pennellii* F2 populations. Theor Appl Genet, 99: 254-271.

Habier D, Fernando R L, Kizilkaya K, et al., 2011. Extension of the bayesian alphabet for genomic selection. BMC Bioinformatics, 12: 186.

Hagassou D, Francia E, Ronga D, et al., 2019. Blossom end-rot in tomato (*Solanum lycopersicum* L.): A multi-disciplinary overview of inducing factors and control strategies. Sci Hortic (Amsterdam), 249: 49-58.

参 考 文 献

Haggard J E, Johnson E B, St Clair D A, 2013. Linkage relationships among multiple QTL for horticultural traits and late blight (*P. infestans*) resistance on chromosome 5 introgressed from wild tomato *solanum habrochaites*. G3: GenesGenomes | Genet, 3: 2131-2146.

Halperin E, Stephan D A, 2009. SNP imputation in association studies. Nat Biotechnol, 27: 349-351.

Hamilton J P, Sim S C, Stoffel K, et al., 2012. Single nucleotide polymorphism discovery in cultivated tomato via sequencing by synthesis. Plant Genome J, 5: 17.

Han P, Lavoir A V, Le Bot J, et al., 2015. Nitrogen and water availability to tomato plants triggers bottom-up effects on the leafminer Tuta absoluta. Sci Rep, 4: 4455.

Hanson P M, Yang R, Wu J, et al., 2004. Variation for antioxidant activity and antioxidants in tomato. J Amer Soc Hort Sci, 129: 704-711.

Hanssen I M, Thomma B, 2010. Pepino mosaic virus: a successful pathogen that rapidly evolved from emerging to endemic in tomato crops. Mol Plant Pathol, 11: 179-189.

Hanssens J, de Swaef T, Steppe K, 2015. High light decreases xylem contribution to fruit growth in tomato. Plant Cell Environ, 38: 487-498.

Haseneyer G, Schmutzer T, Seidel M, et al., 2011. From RNA-seq to large-scale genotyping—genomics resources for rye (*Secale cereale* L.). BMC Plant Biol, 11: 131.

Hayashi T, Iwata H, 2010. EM algorithm for Bayesian estimation of genomic breeding values. BMC Genet, 11: 3.

He S, Schulthess A W, Mirdita V, et al., 2016. Genomic selection in a commercial winter wheat population. Theor Appl Genet, 129: 641-651.

He Y, 2012. Chromatin regulation of flowering. Trends Plant Sci, 17: 556-562.

Hepler P K, 2005. Calcium: a central regulator of plant growth and development. Plant Cell, 17: 2142-2155.

Heslot N, Yang H P, Sorrells M E, et al., 2012. Genomic selection in plant breeding: a comparison of models. Crop Sci, 52: 146-160.

Hess M, Druet T, Hess A, et al., 2017. Fixed-length haplotypes can improve genomic prediction accuracy in an admixed dairy cattle population. Genet Sel Evol, 49: 54.

Hill M, Tran N, 2018. MicroRNAs regulating microRNAs in cancer. Trends Cancer, 4: 465-468.

Hirschi K D, 2004. The calcium conundrum. Both versatile nutrient and specific signal. Plant Physiol, 136: 2438-2442.

Ho L C, 1996. The mechanism of assimilate partitioning and carbohydrate compartmentation in fruit in relation to the quality and yield of tomato. J Exp Bot, 47: 1239-1243.

Hobson G, Grierson D, 1993. Tomato. In: Biochemistry of fruit ripening. Dordrecht:

Springer.

Hobson G E, Bedford L, 1989. The composition of cherry tomatoes and its relation to consumer acceptability. JHort Sci, 64: 321-329.

Hospital F, Charcosset A, 1997. Marker-assisted introgression of quantitative trait loci. Genetics, 147: 1469-1485.

Hospital F, Chevalet C, Mulsant P, 1992. Using markers in gene introgression breeding programs. Genetics, 132: 1199-1210.

How K A, Boureau L, Stammitti-Bert L, et al., 2010. Functional analysis of SlEZ1 a tomato Enhancer of zeste [E (z)] gene demonstrates a role in flower development. Plant Mol Biol, 74: 201-213.

Huang B E, George A W, Forrest K L, et al., 2012. A multiparent advanced generation inter-cross population for genetic analysis in wheat. Plant Biotechnol J, 10: 826-839.

Huang W J, Liu H K, McCormick S, et al., 2014. Tomato pistil factor STIG1 promotes in vivo pollen tube growth by binding to phosphatidylinositol 3-phosphate and the extracellular domain of the pollen receptor kinase LePRK2. Plant Cell, 26 (6): 2505-2523.

Huang Z, Van Der Knaap E, 2011. Tomato fruit weight 11.3 maps close to fasciated on the bottom of chromosome 11. Theor Appl Genet, 123: 465-474.

Hutton S F, Scott J W, Yang W C, et al., 2010. Identification of QTL associated with resistance to bacterial spot race T4 in tomato. Theor Appl Genet, 121: 1275-1287.

Ishibashi K, Masuda K, Naito S, et al., 2007. An inhibitor of viral RNA replication is encoded by a plant resistance gene. In: Proceedings of the national academy of sciences of the United States of America, 104: 13833-13838.

Isidro J, Jannink J L, Akdemir D, et al., 2015. Training set optimization under population structure in genomic selection. Theor Appl Genet, 128: 145-158.

Islam M N, Hasanuzzaman A T M, Zhang Z F, et al., 2017. High level of nitrogen makes tomato plants releasing less volatiles and attracting more Bemisia tabaci (Hemiptera: Aleyrodidae). Front Plant Sci, 8: 466.

Ito Y, Nishizawa-Yokoi A, Endo M, et al., 2017. Re-evaluation of the rin mutation and the role of RIN in the induction of tomato ripening. Nat Plants, 3 (11): 866-874.

Iwata H, Jannink J L, 2010. Marker genotype imputation in a low-marker-density panel with a highmarker-density reference panel: accuracy evaluation in barley breeding lines. Crop Sci, 50: 1269-1278.

Janse J, Schols M, 1995. Une préférence pour un goût sucré et non farineux. Groenten Fruit, 26: 16-17.

Jatoi S A, Fujimura T, Yamanaka S, et al., 2008. Potential loss of unique genetic diversity in tomato landraces by genetic colonization of modern cultivars at a non-center of origin. Plant Breed, 127: 189-196.

Jha U C, Bohra A, Jha R, 2017. Breeding approaches and genomics technologies to increase crop yield under low-temperature stress. Plant Cell Rep, 36: 1-35.

Jiménez-Gómez J M, Alonso-Blanco C, Borja A, et al., 2007. Quantitative genetic analysis of flowering time in tomato. Genome, 50: 303-315.

Johnstone P R, Hartz T K, LeStrange M, et al., 2005. Managing fruit soluble solids with late-season deficit irrigation in drip-irrigated processing tomato production. HortScience, 40: 1857-1861.

Jonas E, de Koning D J, 2013. Does genomic selection have a future in plant breeding? Trends Biotechnol, 31: 497-504.

Jones J B, 1986. Survival of *Xanthomonas campestris* pv. *vesicatoria* in Florida on tomato crop residue, weeds, seeds, and volunteer tomato plants. Phytopathology, 76: 430.

Jones J W, Dayan E, Allen L H, et al., 1991. A dynamic tomato growth and yield model (Tomgro). Am Soc Agri Eng, 34: 663-672.

Jones D A, Thomas C M, Hammondkosack K E, et al., 1994. Isolation of the tomato *cf-9* gene for resistance to *Cladosporium fulvum* by transposon tagging. Science, 266: 789-793.

Kabelka E, Franchino B, Francis D M, 2002. Two loci from *Lycopersicon hirsutum* LA407 confer resistance to strains of *Clavibacter michiganensis* subsp. *michiganensis*. Phytopathology, 92: 504-510.

Kamal H M, Takashina T, Egashira H, et al., 2001. Introduction of aromatic fragrance into cultivated tomato from the "*peruvianum* complex". Plant Breed, 120: 179-181.

Kang B C, Yeam I, Li H X, et al., 2007. Ectopic expression of a recessive resistance gene generates dominant potyvirus resistance in plants. Plant Biotechnol J, 5: 526-536.

Karimi Z, Sargolzaei M, Robinson J A B, et al., 2018. Assessing haplotype-based models for genomic evaluation in Holstein cattle. Can J Sci, 1-10.

Karlova R, Van Haarst J C, Maliepaard C, et al., 2013. Identification of microRNA targets in tomato fruit development using high-throughput sequencing and degradome analysis. J Exp Bot, 64: 1863-1878.

Kazmi R H, Khan N, Willems L A J, et al., 2012. Complex genetics controls natural variation among seed quality phenotypes in a recombinant inbred population of an interspecific cross between *Solanum lycopersicum* × *Solanum pimpinellifolium*. Plant Cell Environ, 35: 929-951.

Keller M, Simm S, 2018. The coupling of transcriptome and proteome adaptation during development and heat stress response of tomato pollen. BMC Genom, 19: 447.

Kenchanmane R S K, Barnes A C, Schnable J C, et al., 2018. Low-temperature tolerance in land plants: are transcript and membrane responses conserved? Plant Sci, 276: 73-86.

Kimbara J, Ohyama A, Chikano H, et al., 2018. QTL mapping of fruit nutritional and flavor components in tomato (*Solanum lycopersicum*) using genome-wide SSR markers and recombinant inbred lines (RILs) from an intra-specific cross. Euphytica, 214: 210.

King S R, Davis A R, Zhang X, et al., 2010. Genetics, breeding and selection of rootstocks for Solanaceae and Cucurbitaceae. SciHortic (Amsterdam), 127: 106-111.

Kinkade M P, Foolad M R, 2013. Validation and fine mapping of lyc12.1, a QTL for increased tomato fruit lycopene content. Theor Appl Genet, 126: 2163-2175.

Kissoudis C, Chowdhury R, Van Heusden S, et al., 2015. Combined biotic and abiotic stress resistance in tomato. Euphytica, 202: 317-332.

Klay I, Gouia S, Liu M, et al., 2018. Ethylene Response Factors (ERF) are differentially regulated by different abiotic stress types in tomato plants. Plant Sci, 274: 137-145.

Klee H J, 2010. Improving the flavor of fresh fruits: genomics, biochemistry, and biotechnology. New Phytol, 187: 44-56.

Klee H J, 2013. Purple tomatoes: Longer lasting, less disease, and better for you. Curr Biol, 23: R520-R521.

Klee H J, Tieman D M, 2013. Genetic challenges of flavor improvement in tomato. Trends Genet, 29: 257-262.

Klee H J, Tieman D M, 2018. The genetics of fruit flavour preferences. Nat Rev Genet, 19: 347-356.

Klein R J, Zeiss C, Chew E Y, et al., 2005. Complement factor H polymorphism in age-related macular degeneration. Science, 308: 385-389.

Kooke R, Kruijer W, Bours R, et al., 2016. Genome-wide association mapping and genomic prediction elucidate the genetic architecture of morphological traits in arabidopsis. Plant Physiol, 170: 2187-2203.

Korte A, Vilhjálmsson B J, Segura V, et al., 2012. A mixed-model approach for genome-wide association studies of correlated traits in structured populations. Nat Genet, 44: 1066-1071.

Kover P X, Valdar W, Trakalo J, et al., 2009. A multiparent advanced generation inter-cross to fine-map quantitative traits in Arabidopsis thaliana. PLoS Genet, 5: e1000551.

Kramer M, Sanders R, Bolkan H, et al., 1992. Postharvest evaluation of transgenic tomatoes with reduced levels of polygalacturonase: processing, firmness and disease resistance. Postharv Biol Technol, 1 (3): 241-255.

Krieger U, Lippman Z B, Zamir D, 2010. The flowering gene *SINGLE FLOWER TRUSS* drives heterosis for yield in tomato. Nat Genet, 42: 459-463.

Kropff M J, Haverkort A J, Aggarwal P K, et al., 1995. Using systems approaches to design and evaluate ideotypes for specific environments. Dordrecht: Kluwer Academic Publisher.

Kusmec A, Srinivasan S, Nettleton D, et al., 2017. Distinct genetic architectures for phenotype means and plasticities in Zea mays. Nat Plants, 3: 715-723.

Kyriacou M C, Rouphael Y, Colla G, et al., 2017. Vegetable grafting: the implications of a growing agronomic imperative for vegetable fruit quality and nutritive value. Front Plant Sci, 8: 741.

Lanfermeijer F C, Warmink J, Hille J, 2005. The products of the broken *Tm-2* and the durable *Tm-2*（2）resistance genes from tomato differ in four amino acids. J Exp Bot, 56: 2925-2933.

Lang Z, Wang Y, Tang K, et al., 2017. Critical roles of DNA demethylation in the activation of ripening-induced genes and inhibition of ripening-repressed genes in tomato fruit. Proc Natl Acad Sci USA, 114 (22): E4511-E4519.

Larbat R, Olsen K M, Slimestad R, et al., 2012. Influence of repeated short-term nitrogen limitations on leaf phenolics metabolism in tomato. Phytochemistry, 77: 119-128.

Laterrot H, 1996. Twenty-one near isogenic lines in Moneymaker type with different genes for disease resistances. Rep Tomato Genet Coop, 46: 34.

Laterrot H, 2000. Disease resistance in tomato: practical situation. Acta Physiol Plant, 22: 328-331.

Laterrot H, Moretti A, 1989. Linkage between Pto and susceptibility to fenthion. Tomato Genet Coop Rep, 39: 21-22.

Le Nguyen K, Grondin A, Courtois B, et al., 2018. Next-generation sequencing accelerates crop gene discovery. Trends Plant Sci, 24: 263-274.

Le L Q, Lorenz Y, Scheurer S, et al., 2006. Design of tomato fruits with reduced allergenicity by dsRNAi-mediated inhibition of ns-LTP（Lyc e 3）expression. Plant Biotechnol J, 4 (2): 231-242.

Lecompte F, Abro M A, Nicot P C, 2010. Contrasted responses of Botrytis cinerea isolates developing on tomato plants grown under different nitrogen nutrition regimes. Plant Pathol, 59: 891-899.

Lecompte F, Nicot P C, Ripoll J, et al., 2017. Reduced susceptibility of tomato stem to the necrotrophic fungus Botrytis cinerea is associated with a specific adjustment of fructose content in the host sugar pool. Ann Bot, 119: 931-943.

Lecomte L, Saliba-Colombani V, Gautier A, et al., 2004. Fine mapping of QTLs for the fruit architecture and composition in fresh market tomato, on the distal region of the long arm of chromosome 2. Mol Breed, 13: 1-14.

Lecomte L, Duffé P, Buret M, et al., 2004. Marker-assisted introgression of 5 QTLs controlling fruit quality traits into three tomato lines revealed interactions between QTLs and genetic backgrounds. Theor Appl Genet, 109: 658-668.

Lee D R, 1990. A unidirectional water flux model of fruit growth. Can J Bot, 68: 1286-1290.

Lee J M, Oh C S, Yeam I, 2015. Molecular markers for selecting diverse disease resistances in tomato breeding programs. Plant Breed Biotechnol, 3: 308-322.

Lee J T, Prasad V, Yang P T, et al., 2003. Expression of *Arabidopsis* CBF1 regulated by an ABA/stress inducible promoter in transgenic tomato confers stress tolerance without affecting yield. Plant Cell Environ, 26 (7): 1181-1190.

Lee S Y, Luna-Guzman I, Chang S, et al., 1999. Relating descriptive analysis and instrumental texture data of processed diced tomatoes. Food Qual Pref, 10: 447-455.

Letort V, Mahe P, Cournede P H, et al., 2008. Quantitative genetics and functional structural plant growth models: Simulation of quantitative trait loci detection for model parameters and application to potential yield optimization. Ann Bot-London, 101: 1243-1254.

Levin I, Gilboa N, Yeselson E, et al., 2000. Fgr, a major locus that modulates the fructose to glucose ratio in mature tomato fruits. Theor Appl Genet, 100: 256-262.

Li Y M, Gabelman W H, 1990. Inheritance of calcium use efficiency in tomatoes grown under low-calcium stress. J Am Soc Hortic Sci, 115: 835-838.

Li J, Liu L, Bai Y, et al., 2011. Seedling salt tolerance in tomato. Euphytica, 178: 403-414.

Li T, Yang X, Yu Y, et al., 2018. Domestication of wild tomato is accelerated by genome editing. Nat Biotechnol, 36: 1160-1163.

Lin K H, Yeh W L, Chen H M, et al., 2010. Quantitative trait loci influencing fruit-related characteristics of tomato grown in high-temperature conditions. Euphytica, 174: 119-135.

Lin T, Zhu G, Zhang J, et al., 2014. Genomic analyses provide insights into the history of tomato breeding. Nat Genet, 46: 1220-1226.

Lippman Z B, Zamir D, 2007. Heterosis: revisiting the magic. Trends Genet, 23: 60-66.

Liu Z, Alseekh S, Brotman Y, et al., 2016. Identification of a Solanum pennellii chromosome 4 fruit flavor and nutritional quality associated metabolite QTL. Front Plant Sci, 7: 1-15.

Liu H, Genard M, Guichard S, et al., 2007. Model-assisted analysis of tomato fruit growth in relation to carbon and water fluxes. J Exp Bot, 58: 3567-3580.

Liu J, Van Eck J, Cong B, et al., 2002. A new class of regulatory genes underlying the cause of pear-shaped tomato fruit. Proc Natl Acad Sci USA, 99: 13302-13306.

参 考 文 献

Liu H J, Yan J, 2019. Crop genome-wide association study: a harvest of biological relevance. Plant J, 97: 8-18.

Liu H, Yu C, Li H, et al., 2015. Overexpression of SHDHN, a dehydrin gene from *Solanum habrochaites* enhances tolerance to multiple abiotic stresses in tomato. Plant Sci, 231: 198-211.

Liu M, Yu H, Zhao G, et al., 2017. Profiling of drought-responsive microRNA and mRNA in tomato using high-throughput sequencing. BMC Genomics, 18: 481.

Liu Y, Zhou T, Ge H, et al., 2016. SSR mapping of QTLs conferring cold tolerance in an interspecific cross of tomato. Intl J Genom, 2016: 1-6.

Lobit P, Génard M, Soing P, et al., 2006. Modelling malic acid accumulation in fruits: relationships with organic acids, potassium, and temperature. J Exp Bot, 57: 1471-1483.

Lobit P, Génard M, Wu B H, et al., 2003. Modelling citrate metabolism in fruits: responses to growth and temperature. J Exp Bot, 54: 2489-2501.

Luo J, 2015. Metabolite-based genome-wide association studies in plants. Curr Opin Plant Biol, 24: 31-38.

Maayan Y, Pandaranayaka E P J, Srivastava D A, et al., 2018. Using genomic analysis to identify tomato *Tm-2* resistance-breaking mutations and their underlying evolutionary path in a new and emerging tobamovirus. Arch Virol, 163: 1863-1875.

Mackay I J, Bansept-Basler P, Barber T, et al., 2014. An eight-parent multiparent advanced generation inter-cross population for winter-sown wheat: creation, properties, and validation. G3: Genes Genom Genet, 4: 1603-1610.

Malundo T M M, Shewfelt R L, Scott J W, 1995, Flavor quality of fresh tomato (*Lycopersicon esculentum* Mill.) as affected by sugar and acid levels. Postharv BiolTechnol, 6: 103-110.

Mangin B, Rincent R, Rabier C E, et al., 2019. Training set optimization of genomic prediction by means of EthAcc. PLoS ONE, 14: e0205629.

Mangin B, Thoquet P, Olivier J, et al., 1999. Temporal and multiple quantitative trait loci analyses of resistance to bacterial wilt in tomato permit the resolution of linked loci. Genetics, 151: 1165-1172.

Manning K, Tör M, Poole M, et al., 2006. A naturally occurring epigenetic mutation in a gene encoding an SBP-box transcription factor inhibits tomato fruit ripening. Nat Genet, 38: 948-952.

Mao L, Begum D, Chuang H, et al., 2000. JOINTLESS is a MADS-box gene controlling tomato flower abscission zone development. Nature, 406: 910-913.

Marchini J, Howie B, 2010. Genotype imputation for genome-wide association studies. Nat Rev Genet, 11: 499-511.

Marschner H, 1983. General introduction to the mineral nutrition of plants. Berlin: Springer.

Martin G B, Brommonschenkel S H, Chunwongse J, et al., 1993. Map-based cloning of a protein kinase gene conferring disease resistance in tomato. Science, (80) 262: 1432-1436.

Martin G B, Frary A, Wu T, et al., 1994. A member of the tomato Pto gene family confers sensitivity to fenthion resulting in rapid cell death. Plant Cell, 6: 1543-1552.

Martre P, Bertin N, Salon C, et al., 2011. Modelling the size and composition of fruit, grain and seed by process-based simulation models. New Phytolt Tansley Review, 191: 601-618.

Martre P, Quilot-Turion B, Luquet D, et al., 2015. Model-assisted phenotyping and ideotype design. London: Academic Press.

Mazzucato A, Cellini F, Bouzayen M, et al., 2015. A TILLING allele of the tomato *Aux/IAA9* gene offers new insights into fruit set mechanisms and perspectives for breeding seedless tomatoes. Mol Breed, 35: 22.

Mazzucato A, Papa R, Bitocchi E, et al., 2008. Genetic diversity, structure and marker-trait associations in a collection of Italian tomato (*Solanum lycopersicum* L.) landraces. Theor Appl Genet, 116: 657-669.

Mboup M, Fischer I, Lainer H, et al., 2012. Trans-species polymorphism and Allele-Specific expression in the *CBF* gene family of wild tomatoes. Mol Biol Evol, 29: 3641-3652.

McCormick S, Niedermeyer J, Fry J, et al., 1986. Leaf disc transformation of cultivated tomato (*L. esculentum*) using Agrobacterium tumefaciens. Plant Cell Rep, 5 (2): 81-84.

McCouch S R, Wright M H, Tung C W, et al., 2016. Open access resources for genome-wide association mapping in rice. Nat Commun, 7: 10532.

McGlasson W B, Last J H, Shaw K J, et al., 1987. Influence of the non-ripening mutantrin and nor on the aroma of tomato fruits. Hort Science, 22: 632-634.

Megraw M, Baev V, Rusinov V, et al., 2006. MicroRNA promoter element discovery in Arabidopsis. RNA, 12: 1612-1619.

Menda N, Semel Y, Peled D, et al., 2004. Insilico screening of a saturated mutation library of tomato. Plant J, 38: 861-872.

Menda N, Strickler S R, Edwards J D, et al., 2014. Analysis of wild-species introgressions in tomato inbreds uncovers ancestral origins. BMC Plant Biol, 14: 287.

Mendell J T, Olson E N, 2012. MicroRNAs in stress signaling and human disease. Cell, 148: 1172-1187.

Meng C, Yang D, Ma X, et al., 2016. Suppression of tomato *SlNAC1* transcription factor delays fruit ripening. J PlantPhysiol, 193: 88-96.

Meng F J, Xu X Y, Huang F L, et al., 2010. Analysis of genetic diversity in cultivated and wild tomato varieties in Chinese market by RAPD and SSR. Agri Sci China, 9: 1430-1437.

Messeguer R, Ganal M, De Vicente M C, et al., 1991. High resolution RFLP map around the root knot nematode resistance gene (*Mi*) in tomato. Theor Appl Genet, 82: 529-536.

Meuwissen T H E, Hayes B J, Goddard M E, 2001. Prediction of total genetic value using genomewide dense marker maps. Genetics, 157: 1819-1829.

Miller J C, Tanksley S D, 1990. RFLP analysis of phylogenetic relationships and genetic variation in the genus *Lycopersicon*. Theor Appl Genet, 80: 437-448.

Milligan S B, Bodeau J, Yaghoobi J, et al., 1998. The root knot nematode resistance gene *Mi* from tomato is a member of the leucine zipper, nucleotide binding, leucine-rich repeat family of plant genes. Plant Cell, 10: 1307-1319.

Milner S, et al., 2011. Bioactivities of glycoalkaloids and their aglycones from *Solanum* species. J Agri Food Chem, 59: 3454-3484.

Minamikawa M F, Nonaka K, Kaminuma E, et al., 2017. Genome-wide association study and genomic prediction in citrus: Potential of genomics-assisted breeding for fruit quality traits. Sci Rep, 7: 4721.

Minoïa S, Bendahmane A, Piron F, et al., 2010. An induced mutation in tomato *eIF4E* leads to immunity to two potyviruses. PLoS ONE, 5: e11313.

Mirnezhad M, Romero-Gonzalez R R, Leiss K A, et al., 2010. Metabolomic analysis of host plant resistance to thrips in wild and cultivated tomatoes. Phytochem Analys, 21 (1): 110-117.

Mirouze M, Paszkowski J, 2011. Epigenetic contribution to stress adaptation in plants. Curr Opin Plant Biol, 14: 267-274.

Mitchell J, Shennan C, Grattan S, 1991. Developmental-changes in tomato fruit composition in response to water deficit and salinity. Physiol Plant, 83: 177-185.

Mohorianu I, Schwach F, Jing R, et al., 2011. Profiling of short RNAs during fleshy fruit development reveals stage-specific sRNAome expression patterns. Plant J, 67: 232-246.

Molgaard P, Ravn H, 1988. Evolutionary aspects of caffeoyl ester distribution in dicotyledons. Phytochemistry, 27: 2411-2421.

Monforte A J, Asíns M J, Carbonell E A, 1996. Salt tolerance in Lycopersicon species. IV. Efficiency of marker-assisted selection for salt tolerance improvement. Theor Appl Genet, 93: 765-772.

Monforte A J, Asíns M J, Carbonell E A, 1997. Salt tolerance in *Lycopersicon* species VI. Genotype by-salinity interaction in quantitative trait loci detection: constitutive and response QTLs. Theor Appl Genet, 95: 706-713.

Monforte A J, Asíns M J, Carbonell E A, 1997. Salt tolerance in *Lycopersicon* species. V. Does genetic variability at quantitative trait loci affect their analysis? Theor Appl

Genet, 95: 284-293.

Monforte A J, Tanksley S D, 2000. Fine mapping of a quantitative trait locus (QTL) from *Lycopersicon hirsutum* chromosome 1 affecting fruit characteristics and agronomic traits: breaking linkage among QTLs affecting different traits and dissection of heterosis for yield. Theor Appl Genet, 100: 471-479.

Moxon S, Jing R, Szittya G, et al., 2008. Deep sequencing of tomato short RNAs identifies microRNAs targeting genes involved in fruit ripening. Genome Res, 18: 1602-1609.

Mu Q, Huang Z, Chakrabarti M, et al., 2017. Fruit weight is controlled by cell size regulator encoding a novel protein that is expressed in maturing tomato fruits. PLoS Genet, 13: e1006930.

Muir S R, Collins G J, Robinson S, et al., 2001. Overexpression of petunia chalcone isomerase in tomato results in fruit containing increased levels of flavonols. Nat Biotechnol, 19: 470-474.

Mueller L A, Tanksley S D, Giovannoni J J, et al., 2005. The tomato sequencing project, the first cornerstone of the International Solanaceae Project (SOL). Comp Funct Genomics, 6 (3): 153-158.

Müller B S F, Neves L G, de Almeida Filho J E, et al., 2017. Genomic prediction in contrast to a genome-wide association study in explaining heritable variation of complex growth traits in breeding populations of Eucalyptus. BMC Genom, 18: 524.

Munns R, Gilliham M, 2015. Salinity tolerance of crops -what is the cost? New Phytol, 208: 668-673.

Munns R, Tester M, 2008. Mechanisms of salinity tolerance. Annu Rev Plant Biol, 59: 651-681.

Muños S, Ranc N, Botton E, et al., 2011. Increase in tomato locule number is controlled by two single-nucleotide polymorphisms located near WUSCHEL. Plant Physiol, 156 (4): 2244-2254.

Mutshinda C M, Sillanpää M J, 2010. Extended Bayesian LASSO for multiple quantitative trait loci mapping and unobserved phenotype prediction. Genetics, 186: 1067-1075.

Nadeem M, Li J, Wang M, et al., 2018. Unraveling field crops sensitivity to heat stress: mechanisms, approaches, and future prospects. Agronomy, 8: 128.

Nakazato T, Warren D L, Moyle L C, 2010. Ecological and geographic modes of species divergence in wild tomatoes. Amer J Bot, 97: 680-693.

Navarro J M, Flores P, Carvajal M, et al., 2005. Changes in quality and yield of tomato fruit with ammonium, bicarbonate and calcium fertilisation under saline conditions. J Hort Sci Biotechnol, 80: 351-357.

Naves E R, De Ávila Silva L, Sulpice R, et al., 2019. Capsaicinoids: pungency beyond capsicum. Trends Plant Sci, 24: 109-120.

Nawaz M A, Imtiaz M, Kong Q, et al., 2016. Grafting: a technique to modify ion accumulation in horticultural crops. Front Plant Sci, 7: 1457.

Nekrasov V, Wang C, Win J, et al., 2017. Rapid generation of a transgene free powdery mildew resistant tomato by genome deletion. Sci Rep, 7: 482.

Nesbitt T C, Tanksley S D, 2002. Comparative sequencing in the genus *Lycopersicon*: implications for the evolution of fruit size in the domestication of cultivated tomatoes. Genetics, 162: 365-379.

Nombela G, Williamson V M, Muniz M, 2003. The root-knot nematode resistance gene *Mi-1.2* of tomato is responsible for resistance against the whitefly *Bemisia tabaci*. Mol Plant-Microbe Interact, 16: 645-649.

Nuruddin M M, Madramootoo C A, Dodds G T, 2003. Effects of water stress at different growth stages on greenhouse tomato yield and quality. HortScience, 38: 1389-1393.

Ofner I, Lashbrooke J, Pleban T, et al., 2016. *Solanum pennellii* backcross inbred lines (BILs) link small genomic bins with tomato traits. Plant J, 87: 151-160.

Ohlson E W, Foolad M R, 2016. Genetic analysis of resistance to tomato late blight in *Solanum pimpinellifolium* accession PI 163245. Plant Breed, 135: 391-398.

Okabe Y, Asamizu E, Saito T, et al., 2011. Tomato TILLING technology: development of a reverse genetics tool for the efficient isolation of mutants from Micro-Tom mutant libraries. Plant Cell Physiol, 52: 1994-2005.

Okello R C O, Heuvelink E, De Visser P H B, et al., 2015. What drives fruit growth? Funct Plant Biol, 42: 817-827.

Oliver J E, Whitfield A E, 2016. The genus tospovirus: emerging bunyaviruses that threaten food security. In: Enquist L W (ed) Annu Rev Virol, 3: 101-124.

Ongom P O, Ejeta G, 2017. Mating design and genetic structure of a multi-parent advanced generation intercross (MAGIC) population of Sorghum [*Sorghum bicolor* (L.) Moench]. G3 Genes | Genomes | Genetics, 8: 331-341.

Osorio S, Ruan Y L, Fernie A R, 2014. An update on source-to-sink carbon partitioning in tomato. Front Plant Sci, 5: 516.

Ould-Sidi M M, Lescourret F, 2011. Model-based design of innovative cropping systems: state of the art and new prospects. AgronSustain Dev, 31 (3): 571-588.

Overy S A, Walker H J, Malone S, et al., 2004. Application of metabolite profiling to the identification of traits in a population of tomato introgression lines. J Exp Bot, 56: 287-296.

Pailles Y, Ho S, Pires I S, et al., 2017. Genetic diversity and population structure of two tomato species from the Galapagos Islands. Front Plant Sci, 8: 138.

Papadopoulos I, Rendig V V, 1983. Interactive effects of salinity and nitrogen on growth and yield of tomato plants. Plant Soil, 73: 47-57.

Paran I, Goldman I, Tanksley S D, et al., 1995. Recombinant inbred lines for genetic mapping in tomato. Theor Appl Genet, 90: 542-548.

Park T, Casella G, 2008. The Bayesian lasso. J Amer Stat Assoc, 103: 681-686.

Park Y H, West M A, St Clair D A, 2004. Evaluation of AFLPs for germplasm fingerprinting and assessment of genetic diversity in cultivars of tomato (*Lycopersicon esculentum* L.). Genome, 47: 510-518.

Pasaniuc B, Rohland N, McLaren P J, et al., 2012. Extremely low-coverage sequencing and imputation increases power for genome-wide association studies. Nat Genet, 44: 631-635.

De Pascale S, Maggio A, Fogliano V, et al., 2001. Irrigation with saline water improves carotenoids content and antioxidant activity of tomato. J Hort Sci Biotechnol, 76: 447-453.

Pascual L, Desplat N, Huang B E, et al., 2015. Potential of a tomato MAGIC population to decipher the genetic control of quantitative traits and detect causal variants in the resequencing era. Plant Biotechnol J, 13: 565-577.

Patanè C, Cosentino S L, 2010. Effects of soil water deficit on yield and quality of processing tomato under a mediterranean climate. Agri Water Manag, 97: 131-138.

Paterson A H, Damon S, Hewitt J D, et al., 1991. Mendelian factors underlying quantitative traits in tomato: comparison across species, generations, and environments. Genetics, 127: 181-197.

Paterson A H, DeVerna J W, Lanini B, et al., 1990. Fine mapping of quantitative trait loci using selected overlapping recombinant chromosomes, in an interspecies cross of tomato. Genetics, 124: 735-742.

Paterson A H, Lander E S, Hewitt J D, et al., 1988. Resolution of quantitative traits into Mendelian factors by using a complete linkage map of restriction fragment length polymorphisms. Nature, 335: 721-726.

Pattison R J, Csukasi F, Zheng Y, et al., 2015. Comprehensive tissue specific transcriptome analysis reveals distinct regulatory programs during early tomato fruit development. Plant Physiol, 168 (4): 1684-1701.

Pertuzé R A, Ji Y, Chetelat R T, 2003. Comparative linkage map of the *Solanum lycopersicoides* and *S. sitiens* genomes and their differentiation from tomato. Genome, 45: 1003-1012.

Petró-Turza M, 1986. Flavor of tomato and tomato products. Food Rev Intl, 2: 309-351.

Pettigrew W T, 2008. Potassium influences on yield and quality production for maize, wheat, soybean and cotton. Physiol Plant, 133: 670-681.

参 考 文 献

Philouze J, 1991. Description of isogenic lines, except for one, or two, monogenically controlled morphological traits in tomato, *Lycopersicon esculentum* Mill. Euphytica, 56: 121-131.

Pillen K, Ganal M W, Tanksley S D, 1996. Construction of a high-resolution genetic map and YACcontigs in the tomato *Tm-2a* region. Theor Appl Genet, 93: 228-233.

Piron F, Nicolai M, Minoia S, et al., 2010. An induced mutation in tomato eIF4E leads to immunity to two potyviruses. PloS One, 5.

Pnueli L, Carmel-Goren L, Hareven D, et al., 1998. The *SELF-PRUNING* gene of tomato regulates vegetative to reproductive switching of sympodial meristems and is the ortholog of CEN and TFL1. Development, 125 (11): 1979-1989.

Poiroux-Gonord F, Bidel L P R, Fanciullino A L, et al., 2010. Health benefits of vitamins and secondary metabolites of fruits and vegetables and prospects to increase their concentrations by agronomic approaches. J Agri Food Chem, 58: 12065-12082.

Poland J A, Balint-Kurti P J, Wisser R J, et al., 2009. Shades of gray: the world of quantitative disease resistance. Trends Plant Sci, 14: 21-29.

Prudent M, Lecomte A, Bouchet J P, et al., 2011. Combining ecophysiological modelling and quantitative trait loci analysis to identify key elementary processes underlying tomato fruit sugar concentration. J Exp Bot, 62: 907-919.

Qi L S, Larson M H, Gilbert L A, et al., 2013. Repurposing *CRISPR* as an RNA-guided platform for sequence-specific control of gene expression. Cell, 152 (5): 1173-1183.

Quadrana L, Almeida J, Asís R, et al., 2014. Natural occurring epialleles determine vitamin E accumulation in tomato fruits. Nat Commun, 5: 4027.

Quilot B, Kervella J, Genard M, et al., 2005. Analysing the genetic control of peach fruit quality through an ecophysiological model combined with a QTL approach. J Exp Bot, 56: 3083-3092.

Quilot-Turion B, Ould-Sidi M M, Kadrani A, et al., 2012. Optimization of parameters of the 'Virtual Fruit' model to design peach genotype for sustainable production systems. Eur J Agron, 42: 34-48.

Rached M, Pierre B, Yves G, et al., 2018. Differences in blossom-end rot resistance in tomato cultivars is associated with total ascorbate rather than calcium concentration in the distal end part of fruits per se Hortic J, 87: 372-381.

Rajasekaran L R, Aspinall D, Paleg L G, 2000. Physiological mechanism of tolerance of *Lycopersicon* spp. exposed to salt stress. Can J Plant Sci, 80: 151-159.

Rajewsky N, 2006. MicroRNA target predictions in animals. Nat Genet, 38: S8-S13.

Rambla J L, Tikunov Y M, Monforte A J, et al., 2014. The expanded tomato fruit volatile landscape. J Exp Bot, 65: 4613-4623.

Ramstein G P, Jensen S E, Buckler E S, 2018. Breaking the curse of dimensionality to identify causal variants in Breeding 4. Theor Appl Genet, 132: 559-567.

Ranc N, Muños S, Xu J, et al., 2012. Genome-wide association mapping in tomato (*Solanum lycopersicum*) is possible using genome admixture of *Solanum lycopersicum* var. *cerasiforme*. G3: Genes Genomes Genet, 2: 853-864.

Ranjan A, Budke J M, Rowland S D, et al., 2016. eQTL regulating transcript levels associated with diverse biological processes in tomato. Plant Physiol, 172 (1): 328-340.

Rao E S, Kadirvel P, Symonds R C, et al., 2013. Relationship between survival and yield related traits in *Solanum pimpinellifolium* under salt stress. Euphytica, 190: 215-228.

Rasmussen S, Barah P, Suarez-Rodriguez M C, et al., 2013. Transcriptome responses to combinations of stresses in arabidopsis. Plant Physiol, 161: 1783-1794.

Rengel Z, 1992. The role of calcium in salt toxicity. Plant Cell Environ, 15: 625-663.

Reymond M, Muller B, Leonardi A, et al., 2003. Combining quantitative trait loci analysis and an ecophysiological model to analyze the genetic variability of the responses of maize leaf growth to temperature and water deficit. Plant Physiol, 131: 664-675.

Rick C M, Chetelat R T, 1995. Utilization of related wild species for tomato improvement. In: FernandezMunoz R, Cuartero J, GomezGuillamon ML (eds) First international symposium on solanaceae for fresh market. Acta Hort, 412: 21-38.

Ripoll J, Urban L, Brunel B, et al., 2016. Water deficit effects on tomato quality depend on fruit developmental stage and genotype. J Plant Physiol, 190: 26-35.

Ripoll J, Urban L, Staudt M, et al., 2014. Water shortage and quality of fleshy fruits—making the most of the unavoidable. J Exp Bot, 65: 4097-4117.

Rivero R M, Mestre T C, Mittler R, et al., 2014. The combined effect of salinity and heat reveals a specific physiological, biochemical and molecular response in tomato plants. Plant Cell Environ, 37: 1059-1073.

Robbins M D, Masud M A T, Panthee D R, et al., 2010. Marker assisted selection for coupling phase resistance to *tomato spotted wilt virus* and *Phytophthora infestans* (Late Blight) in tomato. HortScience, 45: 1424-1428.

Robert V J M, West M A L, Inai S, et al., 2001. Marker assisted introgression of black mold resistance QTL alleles from wild *Lycopersicon cheesmanii* to cultivated tomato (*L. esculentum*) and evaluation of QTL phenotypic effects. Mol Breed, 8: 217-233.

Rodríguez G R, Muños S, Anderson C, et al., 2011. Distribution of SUN, OVATE, LC, and FAS in the tomato germplasm and the relationship to fruit shape diversity. Plant Physiol, 156: 275-285.

Rodríguez-Leal D, Lemmon Z H, Man J, et al., 2017. Engineering quantitative trait variation for crop improvement by genome editing. Cell, 171: 470-480.

Rogers K, Chen X, 2013. Biogenesis, turnover, and mode of action of plant microRNAs. Plant Cell, 25: 2383-2399.

Ronen G, Cohen M, Zamir D, et al., 1999. Regulation of carotenoid biosynthesis during tomato fruit development: expression of the gene for lycopene epsilon-cyclase is down-regulated during ripening and is elevated in the mutant Delta. Plant J, 17: 341-351.

Rosales M A, Rubio-Wilhelmi M M, Castellano R, et al., 2007. Sucrolytic activities in cherry tomato fruits in relation to temperature and solar radiation. Sci Hort (Amsterdam), 113: 244-249.

Rosental L, Perelman A, Nevo N, et al., 2016. Environmental and genetic effects on tomato seed metabolic balance and its association with germination vigor. BMC Genom, 17: 1047.

Rossi M, Goggin F L, Milligan S B, et al., 1998. The nematode resistance gene Mi of tomato confers resistance against the potato aphid. Proc Natl Acad Sci USA, 95: 9750-9754.

Rothan C, Diouf I, Causse M, 2019. Trait discovery and editing in tomato. Plant J, 97: 73-90.

Rousseaux M C, Jones C M, Adams D, et al., 2005. QTL analysis of fruit antioxidants in tomato using *Lycopersicon pennellii* introgression lines. Theor Appl Genet, 111: 1396-1408.

Ruan Y L, Patrick J W, Bouzayen M, et al., 2012. Molecular regulation of seed and fruit set. Trends Plant Sci, 17: 656-665.

Ruffel S, Gallois J L, Lesage M L, et al., 2005. The recessive potyvirus resistance gene *pot-1* is the tomato orthologue of the pepper *pvr2-eIF4E* gene. Mol Genet Genom, 274: 346-353.

Ruggieri V, Francese G, Sacco A, et al., 2014. An association mapping approach to identify favourable alleles for tomato fruit quality breeding. BMC Plant Biol, 14: 1-15.

Sacco A, di Matteo A, Lombardi N, et al., 2013. Quantitative trait loci pyramiding for fruit quality traits in tomato. Mol Breed, 31 (1): 217-222.

Sahu K K, Chattopadhyay D, 2017. Genome-wide sequence variations between wild and cultivated tomato species revisited by whole genome sequence mapping. BMC Genom, 18: 430.

Sainju U M, Dris R, Singh B, 2003. Mineral nutrition of tomato. Food Agri Environ, 1: 176-183.

Saliba-Colombani V, Causse M, Langlois D, et al., 2001. Genetic analysis of organoleptic quality in fresh market tomato: 1. Mapping QTLs for physical and chemical traits. Theor Appl Genet, 102: 259-272.

Sallam A, Martsch R, 2015. Association mapping for frost tolerance using multi-parent advanced generation inter-cross (MAGIC) population in *faba* bean (*Vicia faba* L.). Genetica, 143: 501-514.

Salmeron J M, Oldroyd G E, Rommens C M, et al., 1996. Tomato *Prf* is a member of the leucine-rich repeat class of plant disease resistance genes and lies embedded within the *Pto* Kinase gene cluster. Cell, 86: 123-133.

Sanei M, Chen X, 2015. Mechanisms of microRNA turnover. Curr Opin Plant Biol, 27: 199-206.

Sarlikioti V, De Visser P H B, Buck-Sorlin G H, et al., 2011. How plant architecture affects light absorption and photosynthesis in tomato: towards an ideotype for plant architecture using a functional-structural plant model. Ann Bot, 108 (6): 1065-1073.

Sato S, Tabata S, Hirakawa H, et al., 2012. The tomato genome sequence provides insights into fleshy fruit evolution. Nature, 485: 635-641.

Sauvage C, Rau A, Aichholz C, et al., 2017. Domestication rewired gene expression and nucleotide diversity patterns in tomato. Plant J, 91: 631-645.

Sauvage C, Segura V, Bauchet G, et al., 2014. Genome-wide association in tomato reveals 44 candidate loci for fruit metabolic traits. Plant Physiol, 165: 1120-1132.

Schachtman D P, Shin R, 2007. Nutrient sensing and signaling: NPKS. Annu Rev Plant Biol, 58: 47-69.

Schaffer A A, Levin I, Oguz I, et al., 2000. ADPglucose pyrophosphorylase activity and starch accumulation in immature tomato fruit: the effect of a Lycopersicon hirsutum-derived introgression encoding for the large subunit. Plant Sci, 152: 135-144.

Schauer N, Semel Y, Roessner U, et al., 2006. Comprehensive metabolic profiling and phenotyping of interspecific introgression lines for tomato improvement. Nat Biotechnol, 24: 447-454.

Schauer N, Zamir D, Fernie A R, 2005. Metabolic profiling of leaves and fruit of wild species tomato: a survey of the *Solanum lycopersicum* complex. J Exp Bot, 56: 297-307.

Scheben A, Batley J, Edwards D, 2017. Genotyping-by-sequencing approaches to characterize crop genomes: choosing the right tool for the right application. Plant Biotechnol J, 15: 149-161.

Schijlen E G, De Vos C R, Martens S, et al., 2007. RNA interference silencing of chalcone synthase, the first step in the flavonoid biosynthesis pathway, leads to parthenocarpic tomato fruits. Plant Physiol, 144 (3): 1520-1530.

Scholberg J M S, Locascio S J, 1999. Growth response of snap bean and tomato as affected by salinity and irrigation method. HortScience, 34: 259-264.

Segura V, Vilhjálmsson B J, Platt A, et al., 2012. An efficient multi-locus mixed-model approach for genome-wide association studies in structured populations. Nat Genet, 44: 825-830.

Semel Y, Nissenbaum J, Menda N, et al., 2006. Overdominant quantitative trait loci for yield and fitness in tomato. Proc Natl Acad Sci USA, 103: 12981-12986.

Semel Y, Schauer N, Roessner U, et al., 2007. Metabolite analysis for the comparison of irrigated and non-irrigated field grown tomato of varying genotype. Metabolomics, 3: 289-295.

Shahlaei A, Torabi S, Khosroshahli M, 2014. Efficacy of SCoT and ISSR marekers in assesment of tomato (Lycopersicum esculentum Mill.) genetic diversity. Intl J Biosci, 5: 14-22.

Shalit A, Rozman A, Goldshmidt A, et al., 2009. The flowering hormone florigen functions as a general systemic regulator of growth and termination. Proc Natl Acad Sci USA, 106: 8392-8397.

Shammai A, Petreikov M, Yeselson Y, et al., 2018. Natural genetic variation for expression of a SWEET transporter among wild species of Solanum lycopersicum (tomato) determines the hexose composition of ripening tomato fruit. Plant J, 96: 343-357.

Sharada M S, Kumari A, Pandey A K, et al., 2017. Generation of genetically stable transformants by Agrobacterium using tomato floral buds. Plant Cell Tiss Org Cult, 129 (2): 299-312.

Shimatani Z, Kashojiya S, Takayama M, et al., 2017. Targeted base editing in rice and tomato using a CRISPRCas9 cytidine deaminase fusion. Nature Biotechnol, 35 (5): 441-443.

Shinozaki Y, Nicolas P, Fernandez-Pozo N, et al., 2018. High-resolution spatiotemporal transcriptome mapping of tomato fruit development and ripening. Nat Commun, 9: 364.

Sim S C, Van Deynze A, Stoffel K, et al., 2012. High-density SNP genotyping of tomato (Solanum lycopersicum L.) reveals patterns of genetic variation due to breeding. PLoS One, 7: e45520.

Sim S C, Robbins M D, Van Deynze A, et al., 2010. Population structure and genetic differentiation associated with breeding history and selection in tomato (Solanum lycopersicum L.). Heredity (Edinb), 106: 927-935.

Sim S C, Robbins M D, Wijeratne S, et al., 2015. Association analysis for bacterial spot resistance in a directionally selected complex breeding population of tomato. Phytopathology, 105: 1437-1445.

Simons G, Groenendijk J, Wijbrandi J, et al., 1998. Dissectionof the Fusarium I2 gene cluster in tomato reveals six homologs and one active gene copy. Plant Cell (The), 10:

1055-1068.

Smart C D, Tanksley S D, Mayton H, et al., 2007. Resistance to *Phytophthora infestans* in *Lycopersicon pennellii*. Plant Dis, 91: 1045-1049.

Smirnoff N, Wheeler G L, 2000. Ascorbic acid in plants: biosynthesis and function. Crit RevBiochem Mol Biol, 35: 291-314.

Soyk S, Lemmon Z H, Oved M, et al., 2017. Bypassing negative epistasis on yield in tomato imposed by a domestication gene. Cell, 169: 1142-1155.

Spindel J, Begum H, Akdemir D, et al., 2015. Genomic selection and association mapping in rice (*Oryza sativa*): effect of trait genetic architecture, training population composition, marker number and statistical model on accuracy of rice genomic selection in elite, tropical rice breeding lines. PLoS Genet, 11: 1-25.

Stamova B S, Chetelat R T, 2000. Inheritance and genetic mapping of *cucumber mosaic virus* resistance introgressed from *Lycopersicon chilense* into tomato. TheorAppl Genet, 101: 527-537.

Stevens M A, 1986. Inheritance of tomato fruit quality components. Plant Breed Rev, 4: 273-311.

Stevens M A, Kader A A, Albright M, 1979. Potential for increasing tomato flavor via increased sugar and acid content. J Amer SocHort Sci, 104: 40-42.

Stevens M A, Kader A A, Albright-Holton M, 1977. Intercultivar variation in composition of locular and pericarp portions of fresh market tomatoes. J Amer Soc Hort Sci, 102: 689-692.

Stevens M R, Lamb E M, Rhoads D D, 1995. Mapping the *Sw-5* locus for *tomato spotted wilt virus* resistance in tomatoes using RAPD and RFLP analyses. Theor Appl Genet, 90: 451-456.

Stikic R, Popovic S, Srdic M, et al., 2003. Partial root drying (PRD): a new technique for growing plants that saves water and improves the quality of fruit. Bulg J Plant Physiol, 164-171.

Stricker S H, Köferle A, Beck S, 2017. From profiles to function in epigenomics. Nat Rev Genet, 18: 51-66.

Struik P C, Yin X Y, de Visser P, 2005. Complex quality traits: now time to model. Trends Plant Sci, 10: 513-516.

Suliman-Pollatschek S, Kashkush K, Shats H, et al., 2002. Generation and mapping of AFLP, SSRs and SNPs in *Lycopersicon esculentum*. Cell Mol Biol Lett, 7: 583-597.

Sun J, Poland J A, Mondal S, et al., 2019. High-throughput phenotyping platforms enhance genomic selection for wheat grain yield across populations and cycles in early stage. Theor Appl Genet, 1-16.

Sun X, Gao Y, Li H, et al., 2015. Over-expression of *SlWRKY39* leads to enhanced resistance to multiple stress factors in tomato. J Plant Biol, 58: 52-60.

Suzuki N, Rivero R M, Shulaev V, et al., 2014. Abiotic and biotic stress combinations. New Phytol, 203: 32-43.

Tadmor Y, Fridman E, Gur A, et al., 2002. Identification of malodorous, a wild species allele affecting tomato aroma that was selected against during domestication. J Agri Food Chem, 50: 2005-2009.

Takken F L W, Thomas C M, Joosten M, et al., 1999. A second gene at the tomato *Cf-4* locus confers resistance to *Cladosporium fulvum* through recognition of a novel avirulence determinant. Plant J, 20: 279-288.

Takken F L W, Schipper D, Nijkamp H J J, et al., 1998. Identification and Ds-tagged isolation of a new gene at the *Cf-4* locus of tomato involved in disease resistance to *Cladosporium fulvum* race 5. Plant J, 14: 401-411.

Tam S M, Mhiri C, Vogelaar A, et al., 2005. Comparative analyses of genetic diversities within tomato and pepper collections detected by retrotransposonbased SSAP, AFLP and SSR. Theor Appl Genet, 110: 819-831.

Tanksley S D, 2004. The genetic, developmental, and molecular bases of fruit size in tomato and shape variation. Plant Cell, 16: 181-190.

Tanksley S D, Ganal M W, Prince J P, et al., 1992. High density molecular linkage maps of the tomato and potato genomes. Genetics, 132: 1141-1160.

Tanksley S D, Grandillo S, Fulton T M, et al., 1996. Advanced backcross QTL analysis in a cross between an elite processing line of tomato and its wild relative *L. pimpinellifolium*. Theor Appl Genet, 92: 213-224.

Tanksley S D, Nelson J C, 1996. Advanced backcross QTL analysis: a method for the simultaneous discovery and transfer of valuable QTLs from unadapted germplasm into elite breeding lines. Theor Appl Genet, 92: 191-203.

Tardieu F, 2003. Virtual plants: modelling as a tool for the genomics of tolerance to water deficit. Trends Plant Sci, 8: 9-14.

Tashkandi M, Ali Z, Aljedaani F, et al., 2018. Engineering resistance against *tomato yellow leaf curl virus* via the CRISPR/Cas9 system in tomato. Plant Signal Behav, 13.

Taudt A, Colomé-Tatché M, Johannes F, 2016. Genetic sources of population epigenomic variation. Nat Rev Genet, 17: 319-332.

The 1000 Genomes Project Consortium, 2010. A map of human genome variation frompopulationscale sequencing. Nature, 467: 1061-1073.

The 3000 rice genomes project, 2014. The 3 000 rice genomes project. Gigascience, 3: 7.

The UK10K Consortium, 2015. The UK10K project identifies rare variants in health and disease. Nature, 526: 82-89.

Thoen M P M, Davila O N H, Kloth K J, et al., 2017. Genetic architecture of plant stress resistance: multi-trait genome-wide association mapping. New Phytol, 213: 1346-1362.

Tieman D, Bliss P, McIntyre L M M, et al., 2012. The chemical interactions underlying tomato flavor preferences. Curr Biol, 22: 1035-1039.

Tieman D, Zhu G, Resende M F R, et al., 2017. A chemical genetic roadmap to improved tomato flavor. Science, (80-) 355: 391-394.

Tieman D M, Handa A K, 1994. Reduction in pectin methylesterase activity modifies tissue integrity and cation levels in ripening tomato (*Lycopersicon esculentum* Mill.) fruits. Plant Physiol, 106 (2): 429-36.

Tikunov Y, Lommen A, Vos C H R, et al., 2005. A novel approach for nontargeted data analysis for metabolomics. Large-scale profiling of tomato fruit volatiles. Plant Physiol, 139: 1125-1137.

Tranchida-Lombardo V, Aiese Cigliano R, Anzar I, et al., 2018. Whole-genome re-sequencing of two Italian tomato landraces reveals sequence variations in genes associated with stress tolerance, fruit quality and long shelf-life traits. DNA Res, 25: 149-160.

Tuna A L, Kaya C, Ashraf M, et al., 2007. The effects of calcium sulphate on growth, membrane stability and nutrient uptake of tomato plants grown under salt stress. Environ Exp Bot, 59: 173-178.

Turina M, Kormelink R, Resende R O, 2016. Resistance to tospoviruses in vegetable crops: epidemiological and molecular aspects. In: Leach JE, Lindow S (eds) Annu Rev Phytopathol, 54: 347-371.

Uluisik S, Chapman N H, Smith R, et al., 2016. Genetic improvement of tomato by targeted control of fruit softening. Nature Biotechnol, 34 (9): 950.

Usadel B, Chetelat R, Koren S, et al., 2017. De novo assembly of a new *Solanum pennellii* accession using nanopore sequencing. Plant Cell, 29: 2336-2348.

Vakalounakis D J, Laterrot H, Moretti A, et al., 1997. Linkage between *Frl* (*Fusarium oxysporum* f. sp. *radicis-lycopersici* resistance) and *Tm-2* (*tobacco mosaic virus* resistance-2) loci in tomato (*Lycopersicon esculentum*). Ann Appl Biol, 130: 319-323.

Van Berloo R, Stam P, 1998. Marker-assisted selection in autogamous RIL populations: a simulation study. Theor Appl Genet, 96: 147-154.

Van Berloo R, Stam P, 1999. Comparison between marker-assisted selection and phenotypical selection in a set of Arabidopsis thaliana recombinant inbred lines. Theor Appl Genet, 98: 113-118.

Van Berloo R, Zhu A, Ursem R, et al., 2008. Diversity and linkage disequilibrium analysis within a selected set of cultivated tomatoes. Theor Appl Genet, 117: 89-101.

参 考 文 献

Van der Knaap E, Tanksley S D, 2003. The making of a bell pepper-shaped tomato fruit: identification of loci controlling fruit morphology in Yellow Stuffer tomato. Theor Appl Genet, 107: 139-147.

Van Eeuwijk F A, Bustos-Korts D, Millet E J, et al., 2019. Modelling strategies for assessing and increasing the effectiveness of new phenotyping techniques in plant breeding. Plant Sci, 282: 23-39.

Vargas-Ponce O, Pérez-Álvarez L F, Zamora-Tavares P, et al., 2011. Assessing genetic diversity in mexican husk tomato species. Plant Mol Biol Rep, 29: 733-738.

Veillet F, Perrot L, Chauvin L, et al., 2019. Transgene-free genome editing in tomato and potato plants using Agrobacterium-mediated delivery of a CRISPR/Cas9 cytidine base editor. Intl J Mol Sci, 20 (2): 402.

Venter F, 1977. Solar radiation and vitamin C content of tomato fruits. Acta Hortic, 58: 121-127.

Verkerke W, Janse J, Kersten M, 1998. Instrumental measurement and modelling of tomato fruit taste. Acta Hort, 199-206.

Verlaan M G, Hutton S F, Ibrahem R M, et al., 2013. The *tomato yellow leaf curl virus* resistance genes *Ty-1* and *Ty-3* are allelic and code for DFDGD-Class RNA-dependent RNA polymerases. PLoS Genetics, 9.

Villalta I, Bernet G P, Carbonell E A, et al., 2007. Comparative QTL analysis of salinity tolerance in terms of fruit yield using two *solanum populations* of F7 lines. Theor Appl Genet, 114: 1001-1017.

Víquez-Zamora M, Vosman B, Van De Geest H, et al., 2013. Tomato breeding in the genomics era: insights from a SNP array. BMC Genom, 14: 354.

Vos P, Simons G, Jesse T, et al., 1998. The tomato *Mi-1* gene confers resistance to both root-knot nematodes and potato aphids. Nat Biotechnol, 16: 1365-1369.

Vrebalov J, Ruezinsky D, Padmanabhan V, et al., 2002. A MADS-box gene necessary for fruit ripening at the tomato ripening-inhibitor (rin) locus. Science, (80) 296: 343-346.

Wahid A, Gelani S, Ashraf M, et al., 2007. Heat tolerance in plants: an overview. Environ Exp Bot, 61: 199-223.

Wang D, Salah El-Basyoni I, Stephen Baenziger P, et al., 2012. Prediction of genetic values of quantitative traits with epistatic effects in plant breeding populations. Heredity (Edinb), 109: 313-319.

Wang D R, Agosto-Pérez F J, Chebotarov D, et al., 2018. An imputation platform to enhance integration of rice genetic resources. Nat Commun, 9: 3519.

Wang J F, Ho F I, Truong H T H, et al., 2013. Identification of major QTLs associated with stable resistance of tomato cultivar 'Hawaii 7996' to *Ralstonia solanacearum*. Euphytica, 190: 241-252.

Wang K, Li M, Hakonarson H, 2010. ANNOVAR: Functional annotation of genetic variants from high-throughput sequencing data. Nucleic Acids Res, 38: e164.

Wang L, Song X, Gu L, et al., 2013. NOT2 proteins promote polymerase II-dependent transcription and interact with multiple microRNA biogenesis factors in *Arabidopsis*. Plant Cell, 25: 715-727.

Wang R, Tavano E C D R, Lammers M, et al., 2019. Re-evaluation of transcription factor function in tomato fruit development and ripening with CRISPR/Cas9-mutagenesis. Sci Rep, 8: 1696.

Wang Y, Wu W H, 2015. Genetic approaches for improvement of the crop potassium acquisition and utilization efficiency. Curr Opin Plant Biol, 25: 46-52.

Wang Z, Gerstein M, Snyder M, 2009. RNA-Seq: a revolutionary tool for transcriptomics. Nat Rev Genet, 10: 57-63.

Waters A J, Makarevitch I, Noshay J, et al., 2017. Natural variation for gene expression responses to abiotic stress in maize. Plant J, 89: 706-717.

Wilkins K A, Matthus E, Swarbreck S M, et al., 2016. Calcium-mediated abiotic stress signaling in roots. Front Plant Sci, 7: 1296.

Willits M G, Kramer C M, Prata R T, et al., 2005. Utilization of the genetic resources of wild species to create a nontransgenic high flavonoid tomato. J Agri Food Chem, 53: 1231-1236.

Won S Y, Yumul R E, Chen X, 2014. Small RNAs in plants. Molecular biology. New York: Springer.

Xiao H, Jiang N, Schaffner E, et al., 2008. A retrotransposon-mediated gene duplication underlies morphological variation of tomato fruit. Science, 319: 1527-1530.

Xie Z, Allen E, Fahlgren N, et al., 2005. Expression of *Arabidopsis MIRNA* genes. Plant Physiol, 138: 2145-2154.

Xu J, Driedonks N, Rutten M J M, et al., 2017. Mapping quantitative trait loci for heat tolerance of reproductive traits in tomato (*Solanum lycopersicum*). Mol Breed. 37: 58.

Xu J, Ranc N, Muños S, et al., 2013. Phenotypic diversity and association mapping for fruit quality traits in cultivated tomato and related species. Theor Appl Genet, 126: 567-581.

Xu J, Wolters-Arts M, Mariani C, et al., 2017. Heat stress affects vegetative and reproductive performance and trait correlations in tomato (*Solanum lycopersicum*). Euphytica, 213: 156.

Xu W F, Shi W M, Yan F, 2012. Temporal and tissue-specific expression of tomato *14-3-3* gene family in response to phosphorus deficiency. Pedosphere, 22: 735-745.

参 考 文 献

Yamaguchi H, Ohnishi J, Saito A, et al., 2018. An *NBLRR* gene, *TYNBS1*, is responsible for resistance mediated by the *Ty-2* begomovirus resistance locus of tomato. Theoret Appl Genet, 131: 1345-1362.

Yamamoto E, Matsunaga H, Onogi A, et al., 2016. A simulation-based breeding design that uses whole-genome prediction in tomato. Sci Rep 6: 19454.

Yamamoto E, Matsunaga H, Onogi A, et al., 2017. Efficiency of genomic selection for breeding population design and phenotype prediction in tomato. Heredity (Edinb), 118: 202-209.

Yang D Y, Li M, Ma N N, et al., 2017. Tomato *SlGGP-LIKE* gene participates in plant responses to chilling stress and pathogenic infection. Plant Physiol Biochem, 112: 218-226.

Yang X, Caro M, Hutton S F, et al., 2014. Fine mapping of the *tomato yellow leaf curl virus* resistance gene *Ty-2* on chromosome 11 of tomato. Mol Breed, 34: 749-760.

Yasmeen A, Mirza B, Inayatullah S, et al., 2009. In planta transformation of tomato. Plant Mol Biol Rep, 27 (1): 20-28.

Yin X, Kropff M J, Stam P, 1999. The role of ecophysiological models in QTL analysis: the example of specific leaf area in barley. Heredity, 82: 415-421.

You C, Cui J, Wang H, et al., 2017. Conservation and divergence of small RNA pathways and microRNAs in land plants. Genome Biol, 18: 158.

Young N D, Tanksley S D, 1989. RFLP analysis of the size of chromosomal segments retained around the *Tm-2* locus of tomato during backcross breeding. Theor Appl Genet, 77: 353-359.

Yu B, Bi L, Zheng B, et al., 2008. The FHA domain proteins *DAWDLE* in *Arabidopsis* and *SNIP1* in humans act in small RNA biogenesis. Proc NatlAcad Sci USA, 105: 10073-10078.

Yu Y, Jia T, Chen X, 2017. The 'how' and 'where' of plant microRNAs. New Phytol, 216: 1002-1017.

Zamir D, 2001. Improving plant breeding with exotic genetic libraries. Nat Rev Genet, 2: 3-9.

Zanor M I, Rambla J L, Chaïb J, et al., 2009. Metabolic characterization of loci affecting sensory attributes in tomato allows an assessment of the influence of the levels of primary metabolites and volatile organic contents. J Exp Bot, 60: 2139-2154.

Zhang B, Tieman D M, Chen J, et al., 2016. Loss of tomato flavor quality during chilling is associated with reduced expression of volatile biosynthetic genes and a transient alteration in DNA methylation. Proc Natl Acad Sci USA, 113: 12580-12584.

Zhang C, Liu L, Wang X, et al., 2014. The *Ph-3* gene from *Solanum pimpinellifolium* encodes CC-NBS-LRR protein conferring resistance to *Phytophthora infestans*. Theor Appl Genet, 127: 1353-1364.

Zhang J, Zhao J, Liang Y, et al., 2016. Genome-wide association-mapping for fruit quality traits in tomato. Euphytica, 207: 439-451.

Zhang J, Zhao J, Xu Y, et al., 2015. Genome-wide association mapping for tomato volatiles positively contributing to tomato flavor. Front Plant Sci, 6: 1042.

Zhang S, Xie M, Ren G, et al., 2013. CDC5, a DNA binding protein, positively regulates posttranscriptional processing and/or transcription of primary microRNA transcripts. Proc Natl Acad Sci USA, 110: 17588-17593.

Zhang Y, Butelli E, Alseekh S, et al., 2015. Multi-level engineering facilitates the production of phenylpropanoid compounds in tomato. Nat Commun, 6: 8635.

Zhao C, Liu B, Piao S, et al., 2017. Temperature increase reduces global yields of major crops in four independent estimates. Proc Natl Acad Sci USA, 114: 9326-9331.

Zhao J, Sauvage C, Zhao J, et al., 2019. Meta-analysis of genome-wide association studies provides insights into genetic control of tomato flavor. Nat Commun, 10: 1534.

Zhao X, Liu Y, Liu X, et al., 2018. Comparative transcriptome profiling of two tomato genotypes in response to potassium-deficiency stress. Int J Mol Sci, 19: 2402.

Zhong S, Fei Z, Chen Y, et al., 2013. Single-base resolution methylomes of tomato fruit development reveal epigenome modifications associated with ripening. Nat Biotechnol, 31: 154-159.

Zhou R, Wu Z, Cao X, et al., 2015. Genetic diversity of cultivated and wild tomatoes revealed by morphological traits and SSR markers. Genet Mol Res, 14: 13868-13879.

Zhou R, Yu X, Ottosen C O, et al., 2017. Drought stress had a predominant effect over heat stress on three tomato cultivars subjected to combined stress. BMC Plant Biol, 17: 24.

Zhu G, Gou J, Klee H, et al., 2019. Next-gen approaches to flavor-related metabolism. Ann Rev Plant Biol, 70: 187-212.

Zhu G, Wang S, Huang Z, et al., 2018. Rewiring of the fruit metabolome in tomato breeding. Cell, 172: 249-261.

Zhuang K, Kong F, Zhang S, et al., 2019. Whirly1 enhances tolerance to chilling stress in tomato via protection of photosystem II and regulation of starch degradation. New Phytol, 221: 1998-2012.

Zsögön A, Cermák T, Naves E R, et al., 2018. De novo domestication of wild tomato using genome editing. Nat Biotechnol, 36: 1211-1216.

Zsögön A, Cermak T, Voytas D, et al., 2017. Genome editing as a tool to achieve the crop ideotype and de novo domestication of wild relatives: case study in tomato. Plant Sci, 256:

120-130.

Zuo J, Zhu B, Fu D, et al., 2012. Sculpting the maturation, softening and ethylene pathway: the influences of microRNAs on tomato fruits. BMC Genom, 13: 7.

Zuriaga E, Blanca J, Nuez F. 2009. Classification and phylogenetic relationships in *Solanum* section *Lycopersicon* based on *AFLP* and two nuclear gene sequences. Genet Resour Crop Evol, 56: 663-678.

图书在版编目（CIP）数据

番茄基因组设计育种/（法）玛蒂尔德·科斯等著；梁燕等译．—北京：中国农业出版社，2023.5
书名原文：Genomic Designing of Climate-Smart Vegetable Crops
ISBN 978-7-109-30697-4

Ⅰ.①番… Ⅱ.①玛… ②梁… Ⅲ.①番茄—基因组—作物育种 Ⅳ.①S641.236

中国国家版本馆CIP数据核字（2023）第085977号

First published in English under the title
Genomic Designing for Climate-Smart Tomato,
By Mathilde Causse, Jiantao Zhao, Isidore Diouf, Jiaojiao Wang, Veronique Lefebvre, Bernard Caromel, Michel Génard & Nadia Bertin, https：//doi.org/10.1007/978-3-319-97415-6_2
Selected from Kole, Chitaranjan (eds.): Genomic Designing of Climate-Smart Vegetable Crops.
© Springer Nature Switzerland AG，2020
This edition has been translated and published under permissions from Springer Nature Switzerland AG, part of Springer Nature.

本书内容的任何部分，事先未经出版者书面许可，不得以任何方式或手段复制或刊载。
合同登记号：图字01-2023-2195

FANQIE JIYINZU SHEJI YUZHONG

中国农业出版社出版
地址：北京市朝阳区麦子店街18号楼
邮编：100125
责任编辑：郭晨茜　谢志新
版式设计：王　晨　责任校对：刘丽香
印刷：北京通州皇家印刷厂
版次：2023年5月第1版
印次：2023年5月北京第1次印刷
发行：新华书店北京发行所
开本：700mm×1000mm 1/16
印张：8
字数：160千字
定价：128.00元

版权所有·侵权必究
凡购买本社图书，如有印装质量问题，我社负责调换。
服务电话：010-59195115　010-59194918